Preface

To the Trainee

Welcome to the power industry! This industry includes more than half a million employees who work in a variety of challenging and rewarding occupations. From energy production to transmission and distribution, energy workers fill key roles in this important and growing industry.

According to the Energy Information Administration, electricity consumption is expected to increase by 49 percent by the year 2025—just to keep up with the energy needs of a growing population and economy.

This demand for electricity fuels the need for qualified workers to maintain and operate the generating plants and to service and repair the complex network of wires and transformers that stretch across the country.

Job opportunities in this industry will also increase, as nearly half its technical workforce will reach retirement age within the next ten years. Trained and qualified energy workers will be needed to fill these positions.

As the United States increases its efforts to develop renewable energy resources, such as wind and solar alternatives, energy companies will be on the lookout for qualified workers to install and service these new systems and turbines.

However you look at it, the energy industry is ripe with opportunity for those with the skills, interest, and willingness to learn. If you've started with this course, you're already on your way.

Power Industry Fundamentals

Combined with NCCER's *Core Curriculum*, the *Introduction to the Power Industry* completes *Power Industry Fundamentals*. This course is a prerequisite to more specific training in power generation and transmission and distribution. If you're training through an NCCER-Accredited Training Program Sponsor and successfully pass the module exams and performance tests in this course, you may be eligible for credentialing through NCCER's National Registry. Check with your instructor or local program sponsor to find out. To learn more, go to www.nccer.org or contact us at 1.888.622.3720.

We invite you to visit the NCCER website at www.nccer.org for the latest releases, training information, newsletter, and much more. You can also reference the Contren® product catalog online at www.nccer.org.

Your feedback is welcome. You may email your comments to curriculum@nccer.org or send general comments and inquiries to info@nccer.org.

Contren® Learning Series

The National Center for Construction Education and Research (NCCER) is a not-for-profit 501(c)(3) education foundation established in 1995 by the world's largest and most progressive construction companies and national construction associations. It was founded to address the severe workforce shortage facing the industry and to develop a standardized training process and curricula. Today, NCCER is supported by hundreds of leading construction and maintenance companies, manufacturers, and national associations. The Contren® Learning Series was developed by NCCER in partnership with Pearson Education, Inc., the world's largest educational publisher.

Some features of NCCER's Contren® Learning Series are as follows:

- An industry-proven record of success
- Curricula developed by the industry for the industry
- National standardization providing portability of learned job skills and educational credits
- Compliance with Office of Apprenticeship requirements for related classroom training (CFR 29:29)
- Well-illustrated, up-to-date, and practical information

NCCER also maintains a National Registry that provides transcripts, certificates, and wallet cards to individuals who have successfully completed modules of NCCER's Contren® Learning Series. *Training programs must be delivered by an NCCER Accredited Training Sponsor in order to receive these credentials.*

Contren® Curricula

NCCER's training programs comprise nearly 80 construction, maintenance, pipeline, and utility areas and include skills assessments, safety training, and management education.

Boilermaking
Cabinetmaking
Carpentry
Concrete Finishing
Construction Craft Laborer
Construction Technology
Core Curriculum:
  Introductory Craft Skills
Drywall
Electrical
Electronic Systems Technician
Heating, Ventilating, and
  Air Conditioning
Heavy Equipment Operations
Highway/Heavy Construction
Hydroblasting
Industrial Coating and Lining
  Application Specialist
Industrial Maintenance
  Electrical and
  Instrumentation Technician
Industrial Maintenance
  Mechanic
Instrumentation
Insulating
Ironworking
Masonry
Millwright
Mobile Crane Operations
Painting
Painting, Industrial
Pipefitting
Pipelayer
Plumbing
Reinforcing Ironwork
Rigging
Scaffolding
Sheet Metal
Site Layout
Sprinkler Fitting
Tower Crane Operator
Welding

Green/Sustainable Construction
Your Role in the Green
  Environment
Sustainable Construction
  Supervisor
Introduction to Weatherization
Weatherization Installer

Energy
Introduction to the Power
  Industry
Power Industry Fundamentals
Power Generation Maintenance
  Electrician
Power Generation I&C
  Maintenance Technician
Power Generation Maintenance
  Mechanic
Steam and Gas Turbine
  Technician
Introduction to Solar
  Photovoltaics
Introduction to Wind Energy

Pipeline
Control Center Operations,
  Liquid
Corrosion Control
Electrical and Instrumentation
Field Operations, Liquid
Field Operations, Gas
Maintenance
Mechanical

Safety
Field Safety
Safety Orientation
Safety Technology

Management
Introductory Skills for the
  Crew Leader
Project Management
Project Supervision

Supplemental Titles
Applied Construction Math
Careers in Construction
Tools for Success

Spanish Translations
Basic Rigging
  (Principios Básicos de
  Maniobras)
Carpentry Fundamentals
  (Introducción a la
  Carpintería, Nivel Uno)
Carpentry Forms
  (Formas para Carpintería,
  Nivel Trés)
Conccte Finishing, Lcvcl Onc
  (Acabado de Concreto,
  Nivel Uno)
Core Curriculum:
  Introductory Craft Skills
  (Currículo Básico:
  Habilidades Introductorias
  del Oficio)
Drywall, Level One
  (Paneles de Yeso, Nivel Uno)
Electrical, Level One
  (Electricidad, Nivel Uno)
Field Safety
  (Seguridad de Campo)
Insulating, Level One
  (Aislamiento, Nivel Uno)
Masonry, Level One
  (Albañilería, Nivel Uno)
Pipefitting, Level One
  (Instalación de Tubería
  Industrial, Nivel Uno)
Reinforcing Ironwork, Level One
  (Herreria de Refuerzo,
  Nivel Uno)
Safety Orientation
  (Orientación de Seguridad)
Scaffolding
  (Andamios)
Sprinkler Fitting, Level One
  (Instalación de Rociadores,
  Nivel Uno)

Acknowledgments

This curriculum was revised as a result of the farsightedness and leadership
of the following sponsors:

Arizona Public Services
Arizona Public Services Salt River Project
Baltimore City Community College
Baltimore Gas & Electric
Cianbro Corporation
Constellation Energy
Dunnellon High School Power Generation
    Academy
Fluor Corporation
Gaylor, Inc.

Industrial Management and Training Institute
Mastec North America Energy Service
Mechanical Construction Services, Inc.
Oneonta Job Corps
Progress Energy
Quanta Services, Inc.
South Carolina Electric & Gas
Southern Company
TIC - The Industrial Company
VisonQuest - Academy

This curriculum would not exist were it not for the dedication and unselfish energy of those volunteers who
served on the Authoring Team. A sincere thanks is extended to the following:

Russell Battles
David Brzozowski
Allen Burlison
Scott Englehart
Randall Fabry
Charles George
Ernie Griego
Robert "Bob" Groner
Phillip "Chip" Hammock
Michael Hancock

Joe Holley
Kenneth Scott Huggins
Keith Hunter
Gordon Johnson
Chris Kelley
Mark Lagasse
Darren Marvin
Yolanda McCloud
James McGowan
Jim Mitchem,

Johnny Penrod
Michael A. Roedel
Alton Smith
Shane Smith
Walter Ernie Tomlinson
Tony Vazquez,
Marcel Veronneau
Terry Williams

NCCER Partners

American Fire Sprinkler Association
Associated Builders and Contractors, Inc.
Associated General Contractors of America
Association for Career and Technical Education
Association for Skilled and Technical Sciences
Carolinas AGC, Inc.
Carolinas Electrical Contractors Association
Center for the Improvement of Construction
    Management and Processes
Construction Industry Institute
Construction Users Roundtable
Design Build Institute of America
Merit Contractors Association of Canada
Metal Building Manufacturers Association
NACE International
National Association of Manufacturers
National Association of Minority Contractors
National Association of Women in Construction
National Insulation Association
National Ready Mixed Concrete Association
National Technical Honor Society

National Utility Contractors Association
NAWIC Education Foundation
North American Technician Excellence
Painting & Decorating Contractors of America
Portland Cement Association
SkillsUSA
Steel Erectors Association of America
U.S. Army Corps of Engineers
Women Construction Owners & Executives, USA
University of Florida, M.E. Rinker School of Building
    Construction

Contents

49101-10

Introduction to the Power Industry............. 1.i

This module sets the stage for trainees entering the electrical energy production and distribution field. It describes the many ways in which electricity can be produced, from burning fossil fuels such as coal and natural gas, to harnessing nuclear energy and using renewable energy sources such as wind, geo-thermal, and solar energy. The module explains electrical energy production and its distribution and transmission, and briefly explores the economics involved in energy production and delivery. Finally, the module describes the environmental impacts of producing and distributing electrical power, as well as the methods that are employed to mitigate those impacts. (12.5 hours)

Index I.1

Introduction to the Power Industry

Trainee Guide
First Edition

Prentice Hall

Boston   Columbus   Indianapolis   New York   San Francisco   Upper Saddle River
Amsterdam   Cape Town   Dubai   London   Madrid   Milan   Munich   Paris   Montreal   Toronto
Delhi   Mexico City   Sao Paulo   Sydney   Hong Kong   Seoul   Singapore   Taipei   Tokyo

National Center for Construction Education and Research

President: Don Whyte
Director of Product Development: Daniele Stacey
Power Industry Fundamentals Project Manager: Daniele Stacey
Production Manager: Tim Davis
Quality Assurance Coordinator: Debie Ness
Editor: Rob Richardson
Desktop Publishing Coordinator: James McKay
Production Assistant: Laura Wright

NCCER would like to acknowledge the contract service provider for this curriculum:
Topaz Publications, Syracuse, New York.

This information is general in nature and intended for training purposes only. Actual performance of activities described in this manual requires compliance with all applicable operating, service, maintenance, and safety procedures under the direction of qualified personnel. References in this manual to patented or proprietary devices do not constitute a recommendation of their use.

Prentice Hall
is an imprint of

PEARSON

www.pearsonhighered.com

ISBN 13: 978-0-13-215413-0

Introduction to the Power Industry

49101-10

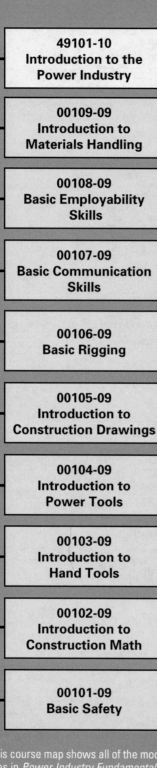

POWER INDUSTRY FUNDAMENTALS

49101-10
Introduction to the Power Industry

00109-09
Introduction to Materials Handling

00108-09
Basic Employability Skills

00107-09
Basic Communication Skills

00106-09
Basic Rigging

00105-09
Introduction to Construction Drawings

00104-09
Introduction to Power Tools

00103-09
Introduction to Hand Tools

00102-09
Introduction to Construction Math

00101-09
Basic Safety

This course map shows all of the modules in *Power Industry Fundamentals*. The suggested training order begins at the bottom and proceeds up. Skill levels increase as you advance on the course map. The local Training Program Sponsor may adjust the training order.

Objectives

When you have completed this module, you will be able to do the following:

1. Define energy and name its sources.
2. Identify the different methods of converting energy into electricity.
3. Explain how electricity is transmitted and distributed.
4. Describe the economics of power generation and distribution.
5. Describe the environmental impacts of producing and distributing electricity and methods used to minimize negative effects.
6. Identify safety considerations associated with the power industry.
7. Identify career opportunities and training requirements in the power industry.
8. Identify concerns that might affect the future of the power industry.

Trade Terms

Alternating current
Bag house
Biogas
Biomass
Blackout
Brownout
Coal gasification
Conductors
Distribution
Efficiency
Electrostatic precipitator
Emissions
Energy
Energy source
Flue gases
Fossil fuel
Generator
Ground

Interconnect
Kinetic energy
Nuclear fission
Nuclear reactor
Organic matter
Photovoltaic
Potential energy
Prime mover
Radioactive
Scrubber
Spent uranium
Stranded costs
Superconductive
Transformer
Transmission
Turbine
Yellowcake

Prerequisites

Before you begin this module, it is recommended that you successfully complete *Core Curriculum*.

POWER INDUSTRY

Contents

Topics to be presented in this module include:

1.0.0 Introduction 1.1
  1.1.0 Electric Power Generation 1.1
  1.2.0 Power Transmission and Distribution 1.3
  1.3.0 Energy 1.4
2.0.0 Fossil-Fuel Power Plants 1.6
  2.1.0 Steam Turbines 1.6
  2.2.0 Combustion Turbines 1.9
3.0.0 Nuclear Power 1.11
  3.1.0 Nuclear Power Plants 1.11
4.0.0 Renewable Energy 1.12
  4.1.0 Biomass 1.12
  4.2.0 Water (Hydropower) 1.13
  4.3.0 Geothermal Energy 1.15
  4.4.0 Wind Energy 1.17
  4.5.0 Solar Energy 1.18
5.0.0 Environmental Controls 1.19
  5.1.0 Electrostatic Precipitators 1.19
  5.2.0 Scrubbers 1.20
  5.3.0 Bag House 1.20
  5.4.0 Selective Catalytic Reduction 1.21
  5.5.0 Water Waste 1.21
  5.6.0 Combustion Waste Products 1.21
6.0.0 Electrical Transmission and Distribution 1.22
  6.1.0 Transmission Subsystem 1.22
    6.1.1 Substations and Switchyards 1.24
  6.2.0 Distribution System 1.26
7.0.0 Careers in the Power Industry 1.27
  7.1.0 Power Plant and Auxiliary Equipment Operators 1.28
  7.2.0 Maintenance 1.28
  7.3.0 Fuel Handling 1.29
  7.4.0 Laboratory Analyst 1.29
  7.5.0 Line Workers 1.29
  7.6.0 Energy Brokers 1.30
8.0.0 Safety Regulations 1.30
  8.1.0 Government Agencies Involved With Safety 1.30
    8.1.1 Occupational Safety and Health Administration 1.30
    8.1.2 Mine Safety and Health 1.31
    8.1.3 American National Standards Institute 1.31
    8.1.4 Environmental Protection Agency 1.31
    8.1.5 Department of Transportation 1.31
    8.1.6 Nuclear Regulatory Commission 1.32
    8.1.7 Transportation Security Administration 1.32

Contents (continued)

9.0.0 Safety Practices.. 1.32
  9.1.0 Policies and Procedures.. 1.32
  9.2.0 Foreign Material Exclusion 1.33
  9.3.0 Grounding 1.34
10.0.0 Power Industry Economics 1.36
  10.1.0 Laws 1.36
    10.1.1 Federal Energy Regulatory Commission 1.37
  10.2.0 Costs 1.37
  10.3.0 Reliability 1.37
11.0.0 The Future of the Power Industry.................................... 1.38
  11.1.0 Environmental Considerations....................................... 1.38
  11.2.0 Future Technology... 1.39
12.0.0 Standardized Training By NCCER.................................... 1.39
  12.1.0 Apprenticeship Training.. 1.40
  12.2.0 Youth Training and Apprenticeship Programs 1.40
Appendix Samples of NCCER Training Credentials 1.49

Figures and Tables

Figure 1 Power-generation station 1.2
Figure 2 Electric transmission lines 1.2
Figure 3 Simple generator 1.3
Figure 4 One cycle of alternating current (AC) 1.3
Figure 5 Simplified diagram of a coal-fired power plant 1.4
Figure 6 Turbine-generator set 1.4
Figure 7 Electrical power distribution 1.5
Figure 8 Coal-burning power plant components 1.7
Figure 9 Turbine rotor 1.8
Figure 10 Steam turbine layout 1.8
Figure 11 Cooling tower 1.8
Figure 12 Basic components of a combustion turbine 1.9
Figure 13 Simple-cycle combustion power plant 1.10
Figure 14 Combined-cycle power plant 1.10
Figure 15 Nuclear power plant 1.12
Figure 16 Hydroelectric power plant 1.13
Figure 17 Water-driven turbine 1.14
Figure 18 Chief Joseph dam 1.14
Figure 19 Old Faithful 1.15
Figure 20 Simplified wet geothermal power plant 1.16
Figure 21 Imperial Valley geothermal power plant 1.16
Figure 22 Wind energy facility in the Northeastern United States 1.17
Figure 23 Sierra SunTower power plant 1.18
Figure 24 Solar panels on public land 1.19
Figure 25 Example of an electrostatic precipitator (dry type) 1.19
Figure 26 Scrubber 1.20
Figure 27 Bag house 1.21
Figure 28 Fuel waste distribution 1.21
Figure 29 Power interconnect regions 1.23
Figure 30 Transmission lines 1.24
Figure 31 Transmission substation transformer 1.24
Figure 32 Subtransmission substation and switchyard 1.25
Figure 33 Distribution substation 1.25
Figure 34 120V/240V power distribution transformers 1.26
Figure 35 Residential power distribution 1.27
Figure 36 Power plant control console 1.28
Figure 37 Coal handling system 1.29
Figure 38 Erecting a power pole 1.29
Figure 39 Power line worker 1.30
Figure 40 FME barrier 1.34
Figure 41 Hard hat lanyard 1.35
Figure 42 FME cover 1.35
Figure 43 Examples of temporary grounding............................ 1.36

Table 1 Typical Transmission, Subtransmission, and
Distribution System Voltages 1.23

1.0.0 INTRODUCTION

It is hard to imagine what our lives would be like without reliable electrical power. Electricity is used to power factories, hospitals, schools, homes, and stores. It also sustains the lives of critically ill hospital patients, and powers streetlights and traffic signals. Scientists knew about and experimented with electricity for many years before the invention of the electric generator and the incandescent light bulb paved the way for the practical delivery and use of electricity.

After inventing the electric light bulb in 1879, Thomas Edison began work on a system for delivering electricity to homes and businesses. His system relied on direct current (DC)—electric current that always flows in one direction. However, DC transmission over long distances proved impractical. Transmitting direct current at the low voltages useful for lighting or motor operation required the use of thick, expensive copper wire. In fact, the service areas of Edison's DC generating stations were limited to about a square mile and mainly served the downtown areas of large cities. While Edison was pioneering DC distribution, electrical engineers in Europe were experimenting with alternating current (AC), which reverses direction at regular intervals.

The American businessman George Westinghouse saw the value in AC. High-voltage AC power could be distributed over longer distances using thinner, less expensive copper wires. Experts theorized that special devices called transformers could be used to increase (step up) and decrease (step down) the voltage level. Increasing the voltage level would allow it to be distributed with minimal loss across a wide region, while reducing the voltage level would enable the current to be used safely in homes and shops once it reached its destination. Westinghouse hired a young electrical engineer named William Stanley, Jr., who developed the first effective transformer and demonstrated the first AC lighting system. Around the same time, the inventor Nikola Tesla filed patents for other devices run by AC. After Westinghouse bought these patents, a full-scale industrial war, known as the current wars, erupted. At stake was whether Edison's direct current or Westinghouse's alternating current would electrify America. Edison claimed that alternating current was extremely dangerous and called for outlawing the high voltages transmitted by AC. Westinghouse countered that transformers safely reduced AC voltages before they entered buildings.

Within ten years, the value of the alternating current system had been convincingly demonstrated. AC proved to be more practical and economical.

Eventually, even Thomas Edison was forced to admit he had been wrong, and General Electric, the company he founded, began building and installing high-voltage AC transmission systems.

1.1.0 Electric Power Generation

Most of the electricity used in the United States is produced by public utilities at power generating plants such as the one shown in *Figure 1*. These plants generate the electricity, which is then distributed to customers using high-voltage transmission lines (*Figure 2*).

The basic purpose of electrical power-generation is to transform an energy source into electrical energy. The energy source is the fuel, such as coal, oil, natural gas, biogas, uranium, wood, wind, or solar. The source will either heat water to make steam or, like wind and solar energy, be directly converted into another form of energy. In the United States, most power generating plants burn coal. Any fuel that can be used to heat water can be used to power a steam turbine power plant. Not all of these fuels are practical, however.

Some of the factors that must be considered include safety, cost, availability, and efficiency. The term *efficiency* refers to how much of the energy released from the source is used to do work.

In most U.S. power plants, a boiler is used to produce steam. The energy source—coal, oil, or natural gas—is burned to heat water in the boiler to a temperature high enough to produce steam. The boiler pressurizes the steam, which is then used to drive a steam turbine.

While the principles behind converting an energy source to electricity are simple, applying them in real life is complicated. Every stage of power generation is carefully regulated to ensure that the output power is consistent, the system runs at optimum efficiency, and any byproducts are disposed of properly.

If the transformer is the key to efficient power transmission and distribution, the generator is the key to power generation. In its simplest form, an electric generator can be represented by a loop of wire rotating in the magnetic field created between the positive and negative poles of a magnet (*Figure 3*). As the conductor (called the rotor) turns, an electric current is induced into it by the

On-Site

Efficiency

Some old coal-burning power plants use only 30 percent of the energy released from coal to generate electricity. Most of the wasted energy is in the form of heat. Power plants that can supply both electricity and heat to buildings are called cogeneration plants. The most efficient cogeneration coal-burning power plant in the world is the Avedore Power Station in Denmark. It uses more than 90 percent of the energy released from coal for electricity and heating.

101F02.EPS

Figure 2 Electric transmission lines.

101F01.EPS

Figure 1 Power generation station.

magnetic field. The magnitude of the induced current varies as the rotor turns, producing an alternating current sine wave (*Figure 4*).

In order for a generator to produce voltage, an external force must be used to turn its rotor. In power-generation plants, that force is provided by a turbine, which is mechanically coupled to the rotor of the generator. The turbine-generator combination is used in all major power-generation facilities. In a coal-fired power plant, for example, coal is burned to heat water in a boiler, converting it into high-pressure steam (see *Figure 5*). The steam is then used to turn the rotor of a steam turbine, which in turn causes the generator rotor to turn.

A turbine (*Figure 6*) has a series of blades mounted on a shaft. Steam, water, or air flows over the blades, causing the shaft to turn. As the turbine turns, the shaft of the generator also turns, producing an AC voltage. In the United States, power generating plants produce voltages with a frequency of 60 Hertz (60 cycles per second). In order to produce a 60-Hz voltage, the basic two-pole generator must rotate at a speed of 7,200 revolutions per minute (rpm). Increasing the number of poles reduces the speed required. This principle is used to compensate for variations in the energy sources for different types of power plants. For example, generators used in hydroelectric plants have many poles to compensate for difficulties in controlling water pressure. The generators at Hoover Dam in Nevada, for example, have 40 poles and turn at 180 rpm to produce 60-Hz power.

1.2.0 Power Transmission and Distribution

The voltage produced by the generator is stepped up to a very high voltage so that it can efficiently travel long distances over the transmission lines (see *Figure 7*). Substation transformers are used to step the voltage down to the levels needed by homes, businesses, and industrial users.

Electrical power must be available to meet the current demand. It cannot be efficiently stored, so it must be produced as it is needed. That means

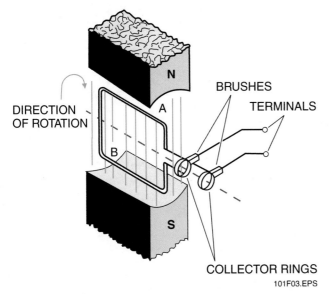

Figure 3 Simple generator.

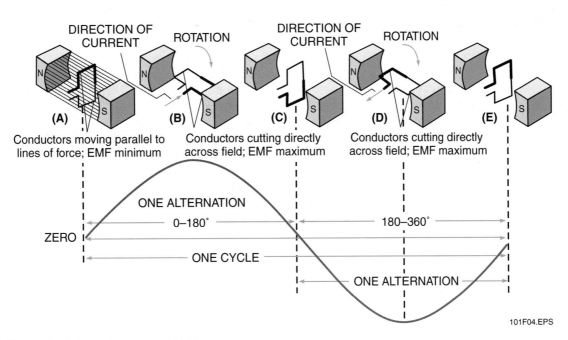

Figure 4 One cycle of alternating current (AC).

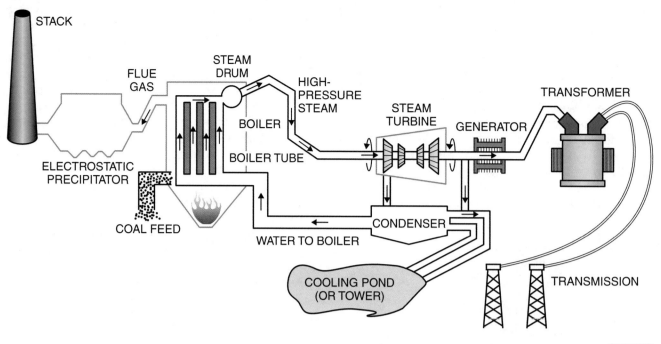

Figure 5 Simplified diagram of a coal-fired power plant.

Figure 6 Turbine-generator set.

plants must be on line at all times to meet demand. The utilities that produce electric power in the U.S. and Canada are interconnected by transmission line, and thus are part of a massive power grid in which power can be exchanged. Within the grid, companies that need additional power to support current demand can buy it from other suppliers on the grid. By the same token, a utility that has surplus power can sell it back to the grid.

Even homeowners and others who use solar, hydro, or wind energy to produce their own power can return surplus power to the grid and obtain credit for it.

1.3.0 Energy

Electricity is a form of energy. Power-generation sources operate on the principle of energy conservation. That is, energy can neither be created nor destroyed; it can only be transformed into other forms of energy. This is known as the first law of thermodynamics. Energy is further described as kinetic energy and potential energy. Kinetic energy is the energy of movement (like a ball rolling down a hill). Potential energy is the capacity for movement (like a ball on a hill before it rolls).

The types of potential energy at work in a power plant are electrical, chemical, thermal, and mechanical. Coal stores chemical energy. When the coal is burned, it undergoes a chemical transformation and releases thermal energy. This energy can be used to boil water, heat a home, or cook food. These are all examples of work.

When coal burns, it is called combustion. Combustion is a chemical reaction between the coal and oxygen in the air that releases heat, so the chemical energy is transformed into thermal energy that is used to produce steam. The steam (thermal energy) is used to turn the rotor of a steam turbine, and is thus converted into mechanical energy. To complete the process, this mechanical energy is converted into electrical energy by the generator. This is a four-way energy conversion: chemical to thermal, to mechanical, to electrical, thus using all forms of potential energy.

Energy can be categorized as renewable or nonrenewable. Nonrenewable sources cannot be replaced in a short period, even with human intervention. Fossil fuels such as coal, oil, and natural gas, along with the uranium used to generate nuclear energy, are examples of nonrenewable resources. Renewable sources can either be easily replaced, or they cannot be exhausted. Wood is a renewable resource that can be replaced. Wind and sunshine are renewable energy sources that cannot be exhausted.

Both renewable and nonrenewable energy sources are used to generate electricity. All fuel sources have advantages and disadvantages. There is no perfect source. Each one must be evaluated to determine its advantages and disadvantages, including cost.

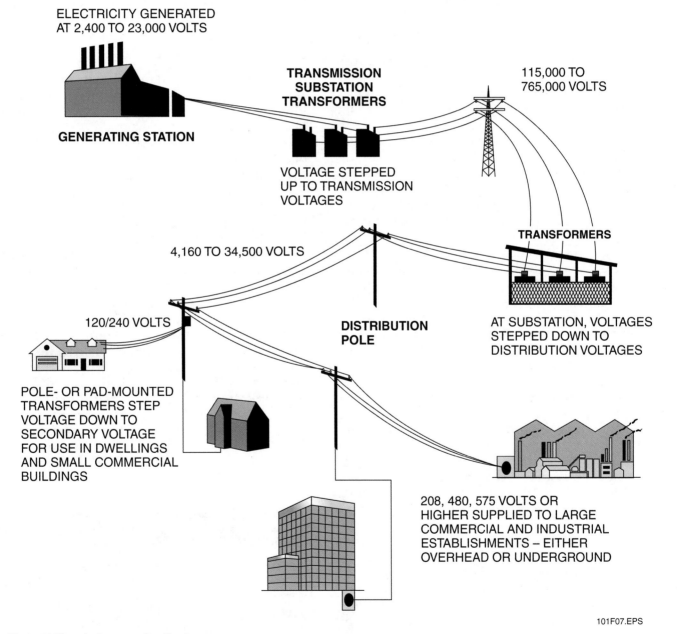

ELECTRICITY GENERATED
AT 2,400 TO 23,000 VOLTS

GENERATING STATION

TRANSMISSION
SUBSTATION
TRANSFORMERS

VOLTAGE STEPPED
UP TO TRANSMISSION
VOLTAGES

115,000 TO
765,000 VOLTS

TRANSFORMERS

4,160 TO 34,500 VOLTS

DISTRIBUTION
POLE

AT SUBSTATION, VOLTAGES
STEPPED DOWN TO
DISTRIBUTION VOLTAGES

120/240 VOLTS

POLE- OR PAD-MOUNTED
TRANSFORMERS STEP
VOLTAGE DOWN TO
SECONDARY VOLTAGE
FOR USE IN DWELLINGS
AND SMALL COMMERCIAL
BUILDINGS

208, 480, 575 VOLTS OR
HIGHER SUPPLIED TO LARGE
COMMERCIAL AND INDUSTRIAL
ESTABLISHMENTS – EITHER
OVERHEAD OR UNDERGROUND

101F07.EPS

Figure 7 Electrical power distribution.

2.0.0 FOSSIL-FUEL POWER PLANTS

Fossil fuels are formed from the remains of dead plants and animals that lived millions of years ago. The dead plants and animals are called organic matter. Fossil fuels are carbon-based and must be burned with oxygen to yield energy.

Coal, petroleum (oil), and natural gas are all examples of fossil fuels. Fuel products made from fossil fuels are also included in this group. They include gasoline, diesel fuel, and propane gas.

Fossil fuels are all found underground. Many coal mines are more than 1,000 feet underground. Oil is often found in pockets 10,000 feet beneath the surface of the Earth or the ocean floor.

The natural gas used as fuel is methane. It is a natural byproduct of coal and oil formation—hence the name natural gas. Natural gas is usually located where coal or oil is present. Wells are dug to tap into pockets of gas just as they are for oil wells. There are small pockets of natural gas scattered around the world, but many of these pockets are too small to be commercially useful.

Among the advantages of fossil fuels is that they are readily available in many areas of the United States and they are relatively inexpensive.

Coal, oil, and gas must be transported to the power plant. Trains and barges are often used to carry coal to power plants, while pipelines are used to carry oil and gas. These systems must be developed and supported so they are safe and reliable.

Coal is the most common type of fuel used in power plants in the United States. Heat, ash, and gases are byproducts of burning coal. Some of these byproducts can be recycled, but when byproducts cannot be reused, they are called waste. Some waste products are harmful and must be captured rather than reused or released into the environment. This topic is covered later in the module.

Figure 8 shows the components of a typical coal-burning power plant. The figure shows the components used to produce electricity, along with the components used to reduce plant emissions.

Coal is pulverized (powdered) before it is burned. Pulverized coal burns with a much higher efficiency than chunks of coal, allowing a greater amount of the energy in the coal to be released.

Coal powder is fed into the burner along with air. The air control subsystem regulates the airflow to burn the coal as fully as possible. Complete combustion of coal cuts down the waste from unburned or incompletely burned coal and increases the amount of heat available to the boiler. Regulating the air and fuel ratios also pre-

vents an explosive atmosphere from occurring in the furnace.

Ash from burned coal, as well as any unburned coal and clumps of impurities that have burned out of the coal, are collected from the burner and from the pollution control systems.

The thermal energy released from burning coal is used to heat water to produce high-pressure steam. Before water can be heated on such a large scale, it must be filtered and conditioned to remove impurities because impurities can be deposited inside the tubes, reducing its efficiency.

High-pressure steam from the boiler is directed across the blades of the steam turbine, causing the turbine rotor to turn. As the turbine rotor turns, so does the rotor of the generator. The output of the generator is electricity that is distributed to consumers via transmission lines.

2.1.0 Steam Turbines

Steam turbines are used in both fossil-fuel and nuclear power plants. A steam turbine is driven by high-pressure steam that is directed onto its rotor blades (*Figure 9*). To understand how this works, consider a simple tea kettle placed on the burner of a stove. As the water in the kettle is heated, the temperature of the water increases. When the water reaches its boiling point (212°F), it will begin to turn into steam, which will be exhausted under pressure through the spout. If you were to place a pinwheel into that exhaust stream, the pressure of the steam would cause the pinwheel to rotate. That is the basic principle of a steam turbine.

A steam turbine consists of blades mounted on a common shaft (*Figure 10*). High-pressure steam is directed onto the turbine blades. As the steam moves along, its pressure is reduced. For that reason, the remaining turbine blades must have increasingly larger surfaces in order to maintain a constant rotation speed. The turbine shaft is mechanically coupled to the rotor of the AC generator. As the shaft turns, the rotor also turns, generating electrical power. In the power industry, whatever turns the generator is called the prime mover; so in this example, the steam turbine is the prime mover.

Did You Know?

It could take a pound of coal to make enough electricity to light a 60-watt bulb for about 15½ hours. One of the most powerful ways to conserve energy is to turn off all lights and electrical appliances when you are not using them.

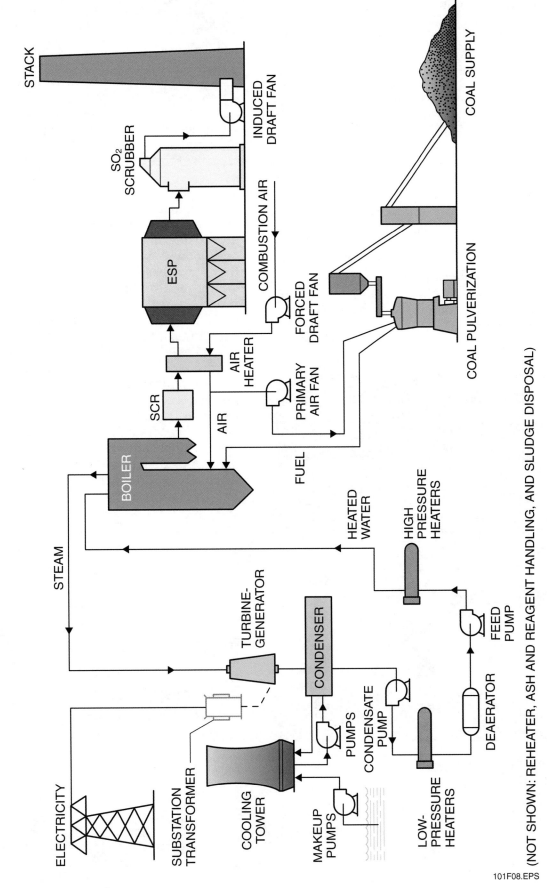

Figure 8 Coal-burning power plant components.

101F08.EPS

Once the energy has been extracted from the steam, the steam is circulated through a condenser, which cools it, causing it to condense back into water. Water from the cooling pond is not mixed with the steam; it is circulated through pipes that run through the steam. As hot steam encounters the cooler pipes inside the condenser, the steam cools and returns to a liquid state. The water is then sent back to the boiler to repeat the cycle.

This recirculation process captures the filtered water, as well as some of the heat that would be lost if the steam was vented. This improves the efficiency of the plant. Waste heat is dissipated through the cooling tower (*Figure 11*).

It takes a great deal of time to bring a coal-burning power plant to full operating capacity, so most of these plants run 24 hours a day, seven days a week. Coal-burning steam-powered plants generate more than half of all of the electricity in the United States.

Although coal is the primary fuel source in U.S. power generating plants, oil and natural gas can also be burned to produce steam. Some plants are able to alternate between these fuels. An advantage of these systems is that the boilers are smaller than those needed to burn coal. In addition, they do not require the coal storage facilities and fuel processing equipment needed for coal-fired boilers. Oil- and gas-fired systems also require less emission-control equipment.

Up until the 1970s, oil was widely used as an energy source for power plants. During the '70s and '80s, oil prices increased sharply, so existing power plants were converted to burn coal. Today less than 15 percent of the electricity in the United States is generated from oil. These plants are expensive to operate and are often used during peak consumption periods, such as hot afternoons when air conditioning use drives up demand.

101F09.EPS

Figure 9 Turbine rotor.

101F11.EPS

Figure 11 Cooling tower.

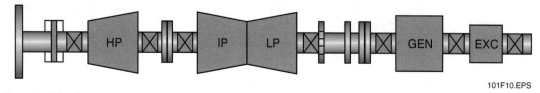

101F10.EPS

Figure 10 Steam turbine layout.

In an oil-powered generation plant, oil is burned with air and the heat released is used to heat water for steam. Oil-burning plants produce the same types of gas emissions as coal, but considerably less ash. In addition, oil that is not completely burned produces a sludge that requires special disposal methods.

2.2.0 Combustion Turbines

Combustion turbines are used to produce electricity in lower-capacity power plants, operating in the 100 to 300 kW range. Plants that rely on large boilers and steam turbines typically produce four to five times that amount. The combustion turbine is self-contained. Unlike the steam turbine, combustion occurs inside of the turbine, producing hot gas that drives the turbine rotor and generator. A jet engine is a type of combustion turbine.

A combustion turbine-generator system (*Figure 12*) consists of the following components:

* Air compressor
* Combustion chamber
* Turbine
* AC generator

In a typical gas-burning power plant, an air compressor supplies compressed air to the tur-

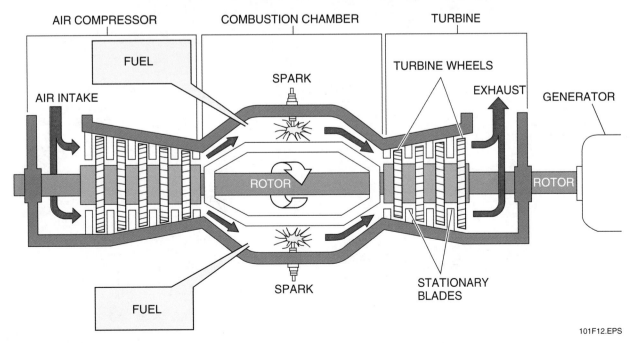

101F12.EPS

Figure 12 Basic components of a combustion turbine.

bine. The compressed air travels to the combustion section (combustor). Gas is also fed into the combustor and the result is a controlled explosion of hot gases (around 2,000°F or 1,093°C). The hot gases are expanded across the blades of the turbine, causing rotation. The rotation of the turbine causes the armature of the AC generator to turn, producing an AC output voltage.

This design is called a simple-cycle combustion turbine (*Figure 13*). The amount of natural gas and air injected into the system is tightly controlled for maximum efficiency. Even so, a simple-cycle combustion turbine has an efficiency of only 20 to 30 percent.

In other power plants, both gas and steam turbines are used. *Figure 14* shows such a plant. The hot gas is captured after it turns the turbine and is used to heat water. The steam from the hot water is used to turn a steam turbine, and is then cooled, condensed, and circulated, in a manner similar to that of a coal-fired unit. This type of power plant is called a combined-cycle power plant. It has an efficiency of almost 60 percent.

In contrast to steam turbine plants, natural gas combustion turbine simple-cycle units (peaking units) can be quickly brought up to full operation. They can easily track loads, so they are often used to supply electricity during peak usage periods. They burn cleaner than coal- or oil-burning plants because they do not produce any sulfur emissions. However, natural gas is typically more expensive than coal.

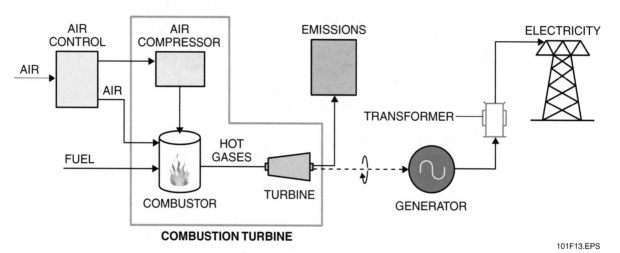

101F13.EPS

Figure 13 Simple-cycle combustion power plant.

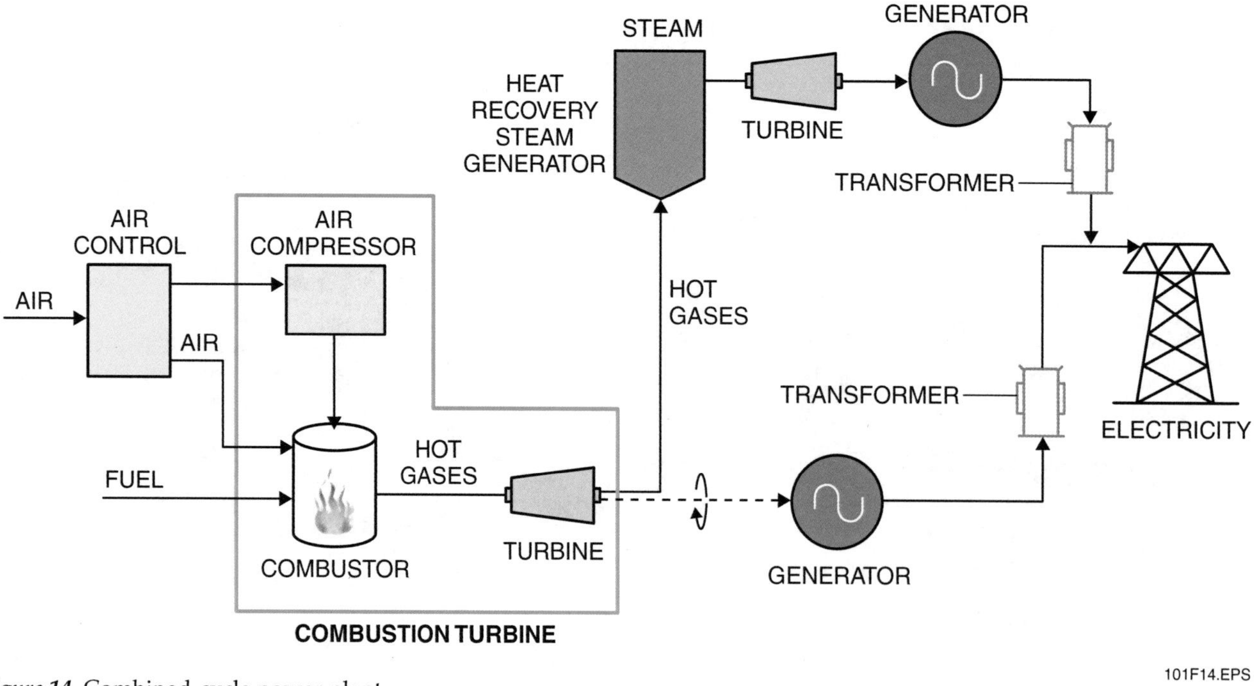

101F14.EPS

Figure 14 Combined-cycle power plant.

3.0.0 NUCLEAR POWER

Nuclear energy sources include uranium and plutonium. Uranium is a radioactive metallic element. It is the most common element used for nuclear power. Plutonium is made from uranium and is used in nuclear weapons. Some countries use plutonium in nuclear power plants, but the United States does not.

Uranium is mined. There are deposits of it scattered around the world, including the United States. When uranium is mined, it is too weak to use as a fuel source in nuclear power plants without being enriched. The raw uranium ore undergoes chemical and physical enrichment processes in which it is compressed into pellets that are stacked inside stainless steel tubes called fuel rods. Enriched uranium is called yellowcake.

Radioactive elements like uranium emit radiation, which is energy moving in the form of particles or waves. While heat, light, radio waves, and microwaves are familiar forms of radiation, a radioactive element releases nuclear radiation, which travels with very high energy. The atomic nucleus of a radioactive element is unstable and emits radiation as it decays. In a nuclear power plant, the release of energy by radioactive elements is controlled and amplified by a process known as nuclear fission. Nuclear fission produces extreme heat, which is used to produce steam to drive turbine-generators.

There are many advantages to using nuclear power to produce electricity. The most important advantage is that it is clean. There are no harmful gases released during nuclear fission and no ash is produced.

The biggest concern with nuclear power is radiation safety. Enriched uranium needs to be handled carefully to prevent its release into the environment. Uranium dust can be inhaled or swallowed. The body excretes most uranium dust, but it can be toxic, especially to the liver and kidneys. Radiation from uranium can be absorbed into the body through the skin, and radioactive particles and waves can pass through cells of the body, causing damage. Even low levels can cause birth defects, and high doses can cause cancer and death.

Disposal of used uranium fuel rods is also a safety issue. As uranium is used, it grows weaker until it is too weak to be used as an energy source. This is called spent uranium. Spent uranium is still radioactive, so it cannot be thrown away. Much of the spent uranium is stored at the power plant that used it, in pools of water that keep it shielded and cooled. It takes decades for the radiation in uranium to decay to safe levels. Currently, there is no known way to speed radiation decay time.

Finally, when some materials are exposed to radiation, they become radioactive themselves. This is what happens in power plants after they have been in operation. Some of the plant equipment, after exposure and contamination, will begin to emit low levels of radiation. The emissions will slowly decay over time as long as the material is not exposed to radiation again. These materials cannot be discarded until the radiation level is very low. For all these reasons, nuclear power plant safety is very tightly regulated.

3.1.0 Nuclear Power Plants

Nuclear fission produces very little waste. A small amount of uranium can be used for one to two years before it needs to be replaced. There are no harmful gases released during nuclear fission and no ash is produced. Typically, nuclear power plants release only insignificant levels of radiation or any nuclear material into the environment. In the immediate vicinity of the nuclear reactor, low levels of radiation are emitted, but the emissions are contained within the facility. Nuclear power plant workers are shielded from exposure to radiation by thick layers of boundary materials, such as steel and concrete.

The operation of a nuclear power plant is very similar to that of coal- and gas-burning steam plants, except that the thermal energy to convert water to steam is obtained by nuclear fission in the reactor (see *Figure 15*). Nuclear power does not produce any harmful gases, ash, or smoke, so there is no need for emission-control systems.

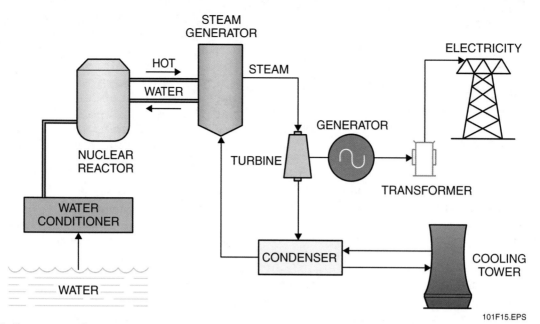

Figure 15 Nuclear power plant.

101F15.EPS

4.0.0 RENEWABLE ENERGY

It is widely recognized that fossil fuels are a diminishing resource. For that reason, a great deal of emphasis is being placed on the use of renewable energy sources, including the following:

- Biomass
- Hydropower
- Geothermal
- Wind
- Solar

While hydropower goes back to the earliest days of electricity generation, use of wind, solar, and biomass resources had only been on a small scale until energy shortages caused by oil embargoes in the late 1900s, along with environmental concerns, began to drive research into more efficient usages of alternative energy sources. Much like the industrial revolution of the 18th century, a chemical and physics revolution is happening now to make renewables not only effective electricity sources but also economically appealing.

Legislation at both state and federal levels requires increasing use of renewable energy throughout the electrical system. With existing and emerging technologies, renewable energies will be a growing part of the nation's energy portfolio and energy independence. Some of these energy sources are not commonly used in the United States, because the technology for converting them into electrical power is not yet mature. The use of these energy sources will become more common in the future. At this time, there is a dramatic increase in the use of wind and solar power.

The use of these resources will reduce our dependence on limited fossil fuels and will increasingly move the U.S. toward energy independence.

4.1.0 Biomass

Biomass fuel comes from recently living sources such as plants, trees, and crops, as well as livestock and human waste, including household garbage. The crops may be specially grown for fuel or may be byproducts of farming, manufacturing, or logging activities. Biomass can be used as a solid fuel, or it can be processed into a liquid or a gas (called biogas). Biomass fuels are almost identical in composition to fossil fuels, except that biomass fuels do not contain sulfur as do coal and oil.

Solid biomass can be burned just like coal, and it can be burned with coal to decrease coal use. Solid biomass can be processed into liquids such as ethanol and bio-diesel. Ethanol is used as a replacement for gasoline in automobiles. Biogas is mostly methane, just like natural gas. Biogas forms naturally when waste decays. Digesters are commercial equipment used to speed up the decay process and capture the methane produced, so they are often used commercially to produce biogas.

There are many advantages to biomass fuels. First, they are renewable. That is, they can be generated from byproducts of other industries as well as from waste material. The sources are readily available in large amounts. Biomass fuels can use waste that would normally go into landfills and release methane gas into the atmosphere as it decays. Some biomass fuel sources are very inexpensive.

There are some disadvantages of using biomass for fuel. Biomass crops are grown on farmland that might otherwise be used for food. Waste from growing food crops was previously tilled back into the farmland as a natural fertilizer. This waste would need to be replaced with chemical fertilizers. As a solid fuel, biomass does not have the heating value of coal, so more needs to be burned to get the same result. Biomass is much more cumbersome than coal or oil and is more costly to ship per unit of energy recovered.

Like fossil fuel, biomass emits carbon dioxide when burned, along with harmful gases. Although this is a disadvantage, remember that when placed in landfills, it decays to produce methane and other harmful gases. Burning biomass as a fuel at least gets work out of material that would otherwise be discarded.

The efficiency of biomass depends on how much water it contains. Fresh biomass contains a lot of water, so it is hard to burn and leaves a lot of residue. Biomass is usually dried to increase its efficiency. It can also be processed into a liquid or gas fuel and then burned. The efficiency of biomass varies with its components and processing.

4.2.0 Water (Hydropower)

Water has long been used as an energy source. For hundreds of years, sawmills and flour mills were built near rivers where fast-moving water provided power to run saws and grinding stones. Today, hydropower electrical plants use the flow of the water to turn turbines to generate electricity

(*Figure 16*). Hydropower generates an estimated 5 to 10 percent of the electricity used in the United States. However, in Canada, 59 percent of the electricity is generated by hydropower.

Hydroelectric power plants need a body of water and some means of making the water flow swiftly, either naturally or in a controlled manner. Most plants dam water into a reservoir to control its flow because it is much more reliable than depending on nature to provide the desired flow rate. The water flows out of the dam onto the blades of the turbines, causing them to turn. Like other types of power plants, the rotors of the turbines are coupled to generators that produce electricity. *Figure 17* shows the type of turbine used in some hydroelectric plants.

There are many advantages of hydroelectric power. First, it does not consume or pollute the water; the water is output from the plant just as it came in. It generates no harmful emissions during operation. It is also very efficient. Well-designed and maintained hydropower plants can have an efficiency of 90 percent. Small hydropower electrical plants are often used to supply electricity to remote areas where it is impractical or difficult to install transmission lines from another source.

Figure 18 shows a picture of the Chief Joseph dam located on the Columbia River in Washington State. It is the second-largest hydroelectric dam in the United States and the largest one operated by the U.S. Army Corps of Engineers. It produces enough electricity to power the city of Seattle.

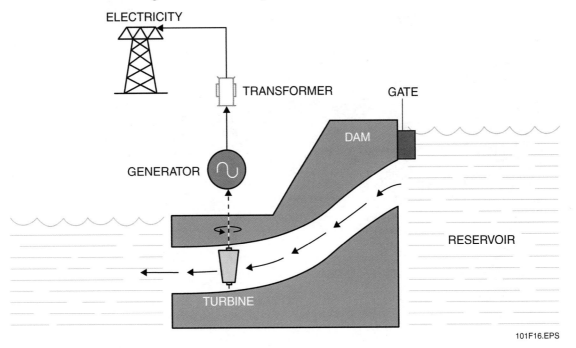

101F16.EPS

Figure 16 Hydroelectric power plant.

101F17.EPS

Figure 17 Water-driven turbine.

There are several disadvantages of hydroelectric power. Building a hydroelectric power plant is a big project. The plant needs a reliable source of water and must be close to the consumers of its electrical output. When people are settled near the proposed site, they must be moved before the plant can be built. Those living and working downstream of the dam and relying on the water for their living, such as farmers, fisheries, and recreation areas, can be adversely affected. Building this type of power plant can have a negative impact on wildlife habitats. Further, stopping the flow of water changes its air content, which can kill organisms in the water. As these organisms decay, they release a small amount of methane gas. This impact is mitigated by introducing air into the turbine, which increases the dissolved oxygen downstream. The construction of dams also can prevent salmon from returning to their breeding grounds. Fish passages/ladders are often created to allow them to pass over the dam.

In addition to the typical maintenance any power plant requires, hydroelectric plants have some special requirements. Sediment in the water, which normally travels with the flow of the water, settles to the bottom of the reservoir, filling it and creating the need for costly dredging. The structural integrity of any dam must be maintained to prevent flooding of nearby areas.

On-Site

Rural Electrification

In the early days of electrical power, electricity became readily available in urban areas as power plants sprung up in those areas. It took much longer for electricity to reach those who lived in remote areas. The solution to that problem was the Rural Electrification Program created by the U.S. Government in the 1930s. Under this program, power plants were built and rivers were dammed to provide the power needed to support industries and make farmers more productive.

101F18.EPS

Figure 18 Chief Joseph dam.

4.3.0 Geothermal Energy

Geothermal energy comes from the hot core of the Earth. Most people think of the Earth as a solid mass, but in reality, it has a core of molten rock called magma. Magma is similar to the molten lava found in volcanoes. Its temperature ranges from 1,300°F to 2,400°F (700°C to 1,300°C).

The surface of the Earth is not hot because there are layers of rock and soil over this molten core. When water finds its way deep into the Earth, it is heated. Sometimes the pressure from the steam is so great that the water and steam are forced to the surface. This is what causes hot springs and geysers. *Figure 19* shows a picture of Old Faithful, a natural geyser located in Yellowstone National Park in Wyoming.

Geothermal energy sources use the natural heat found in the core of the Earth. The best geothermal source for electrical power generation is dry steam from resources with high temperature and relatively low water content. Superheated steam under high pressure is captured through drilled wells and run directly to steam turbines. The largest dry steam field in the world is found in California.

Wet geothermal plants (*Figure 20*) use steam turbines as the prime mover. Brine is pumped under pressure from the resource through a pipeline to a steam separator where some of it is converted (flashed) to steam. The steam, still under pressure, is used to drive a turbine-generator set to produce power. The exhaust steam is condensed and the condensate is re-injected (pumped) back into the resource. Harmful, noncondensible gases from the steam may be captured for disposal, vented under permit rules, or reinjected with the condensate.

Binary geothermal units use the heat from pumped brine to vaporize and superheat various working fluids including refrigerants, ammonia, and hydrocarbons such as isobutane. The brine is pumped from the ground and through a heat exchanger, giving up heat to the working fluid. The cooled brine is pumped back into the resource through reinjection wells. Because the brine is kept under pressure, no mineral or dissolved gas is released. Binary units are often used in resources with lower temperatures and/or volumes of brine than in fields supplying flash plants.

The working fluid is selected based on the temperature and available flow of brine to get maximum efficiency. It is the vaporized and superheated working fluid that drives the turbines.

On-Site

Geothermal Energy

Geothermal energy has few disadvantages. One disadvantage is that depending on where the heat is obtained, gases such as carbon dioxide and methane may be released from the well along with the heat. The amounts of these gases cannot be predicted because each well is unique.

101F19.EPS

Figure 19 Old Faithful.

Turbine exhaust is condensed in air-cooled condensers and fed back into the heating cycle.

Figure 21 shows part of a geothermal power plant located in the Imperial Valley of California. The artesian hot springs, mud pots, and steam vents are too low in thermal energy to be viable for commercial production. These features are found in areas with viable geothermal capacity below the surface. This plant operates turbines using steam from superheated water reservoirs located thousands of feet beneath the Earth's surface.

Dry geothermal energy is heat. It can be used to heat water to generate steam to turn a turbine. This involves drilling deep below the surface to tap into the heat. Current technology is not economically feasible, so dry geothermal energy is

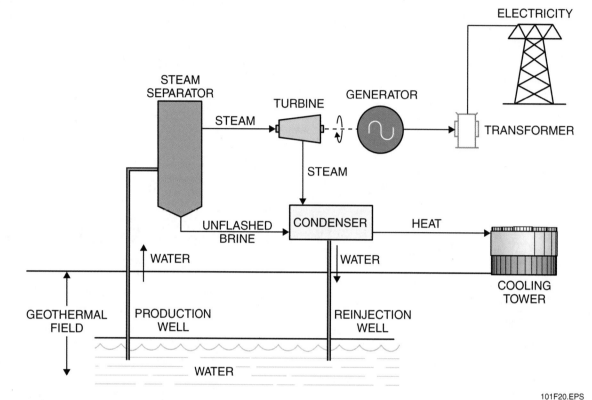

101F20.EPS

Figure 20 Simplified wet geothermal power plant.

101F21.EPS

Figure 21 Imperial Valley geothermal power plant.

not used commercially at this time. Research concerning the effective use of wet and dry geothermal energy to generate electricity is ongoing, so this energy source may be used to a greater degree in the future.

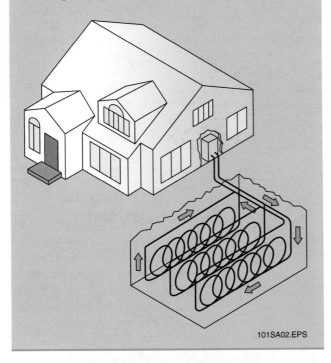
Geothermal energy is classified as a renewable energy source although it is not truly renewable. When heat from the Earth is converted into electricity, it is not replaced. However, the source is so large in some areas that it is viewed as unlimited, and is therefore classified as renewable.

Sometimes water that has been naturally heated by geothermal energy makes it to the Earth's surface as hot water or steam. This source may be used to turn steam turbines without having to dig a well. The steam must be used near the source, which may not be the best place for a power plant.

4.4.0 Wind Energy

Wind has long been used as a source of power. Wind is the result of air movement due to temperature differences between different geographic areas. Wind is considered a renewable resource even though it is generated without human intervention. Large-scale, wind-powered electrical power generation is a fairly new technology in the United States, but other countries have been using it for years.

Wind has the obvious advantage of being readily available in some areas. However, wind is affected by many things, such as mountains, trees, and tall buildings, so not all areas can be used efficiently to produce wind power. When land is devoted to capturing wind for power, it is called a wind farm. One study estimates that only about 6 percent of the land in the United States is suitable for use as wind farms (see *Figure 22*). For this reason, wind farms are sometimes located at sea.

101F22.EPS

Figure 22 Wind energy facility in the Northeastern United States.

Wind power has its disadvantages. For wind farms to be cost effective, the wind must be fairly steady and strong enough to generate enough electricity to offset the cost of building a farm. This translates into wind speeds of 11 to 12 miles per hour on land and 14 to 15 miles per hour at sea. In addition, to some people, wind farms are unattractive and noisy.

4.5.0 Solar Energy

Solar power comes from sunlight, the most powerful energy source known to man. Solar power has long been used as an energy source. Before electricity, solar power was used to dry foods, bleach wood, and heat homes. This energy source is readily available, in at least a limited amount, in all areas of the world. It is renewable with no human intervention and it does not pollute the environment.

There are two types of solar-powered electrical generation. Concentrated solar power (CSP) uses reflective devices to focus the sun's energy onto a collector. The CSP power-generation facility shown in *Figure 23* contains 24,000 sun-tracking mirrors called heliostats. The mirrors reflect sunlight onto a tower-mounted receiver. The reflected energy, in the form of heat, is used to produce steam that is used to drive a turbine-generator set. This facility can produce 5 MW of electricity. The second type of solar power energy generator is a solid-state electronic device called a photovoltaic (PV) cell. These cells are made of silicon combined with doping materials. When sunlight strikes the photovoltaic cell, it produces a DC voltage. This voltage is converted into AC using a device known as an inverter. *Figure 24* shows a power-generation facility that uses arrays of PV cells to generate power.

The advantages of solar energy are obvious. It is renewable without any human intervention and virtually any area can support at least some solar-powered electrical generation. Some people use solar systems to power their homes, thus eliminating the need for power poles and transmission lines. The use of solar panels to heat domestic hot water is very common in the U.S.

Solar power has its disadvantages. Sunlight is not available on a consistent basis. At night, on cloudy days, and when daylight hours are short due to seasonal changes, there is no electrical power generation. However, a device that can store electricity during sunny periods can be used. Also, the manufacture of solar devices is expensive, and solar generation plants require more space than a conventional fossil fuel plant.

GOING GREEN

Timing is Everything

Solar power facilities are not able to produce as much electricity as other types of power plants. However, they are at peak efficiency during the afternoon hours. On a hot day, these facilities can provide the power needed to meet the increased demand created by air conditioning.

101F23.EPS

Figure 23 Sierra SunTower power plant.

101F24.EPS

Figure 24 Solar panels on public land.

5.0.0 ENVIRONMENTAL CONTROLS

Burning fossil fuels produces a large amount of waste products. Many of these waste products cannot be released to the environment. Pollution-control equipment is used to remove and dispose of these pollutants. The following sections introduce the pollution-control equipment commonly used in coal-burning power plants.

Whenever possible, these waste products are captured and/or recycled.

5.1.0 Electrostatic Precipitators

An electrostatic precipitator (ESP) removes ash and other solid particles from flue gases (see *Figure 25*). When pulverized coal fuel is burned, it produces ash. A small amount of the ash is very heavy and falls to the bottom of the burner unit but most of it is light and floats in the flue gas.

Recycling Waste Products

GOING GREEN

The U.S. Environmental Protection Agency (EPA) and a number of other agencies, both private and governmental, sponsor a program to find uses for coal combustion waste. It is called the Coal Combustion Products Partnership (C^2P^2).

This is called fly ash. Approximately 60 to 85 percent of ash is collected from the flue gas.

The electrostatic precipitator removes fly ash particles by negatively charging them with electricity. The negative charge causes the particles to cling to positively charged plates. The particles are then removed from the plates by a mechanism that raps the plates, causing the particles to fall to the bottom of the ESP.

101F25.EPS

Figure 25 Example of an electrostatic precipitator (dry type).

There are dry- and wet-bottom ESPs. In a dry-bottom unit, particles fall into hoppers. They can then be removed through the bottom. In wet-bottom units, the trapped particles are removed by flushing them out with a liquid.

The large volumes of flue gas from the coal means ESPs must be very large. A typical ESP at a power plant can be over 50 feet long, 50 feet wide, and 50 feet high. ESPs can remove up to 99 percent of the ash before the flue gas is discharged from the smoke stack.

5.2.0 Scrubbers

A scrubber is used to remove sulfur dioxide (SO_2) gas from the stack gases before it can be exhausted to the air and create smog. Although there are both wet and dry scrubbers, wet scrubbers are more common. In a wet scrubber, the exhaust gases are sprayed with a slurry (mixture) of water and powdered limestone or lime. Limestone and lime are both forms of calcium. A pair of scrubbers can be seen in *Figure 26*, one scrubber per unit, feeding into the dual chimney stack.

> **NOTE**
> The largest contributors to smog today are vehicle and industrial emissions rather than coal-burning power plants.

The calcium in the slurry combines with the sulfur dioxide particles to form calcium sulfate, which is synthetic gypsum. Gypsum is commonly

used to make drywall board for the construction industry, so some of the gypsum is sold to drywall manufacturers. Other uses for gypsum include the following:

- Construction fill
- Acid neutralizing agent used in mining
- Soil surface stabilizer
- Acid neutralizer for soil

5.3.0 Bag House

A bag house (*Figure 27*) is used to collect ash and solids downstream of a scrubber. A series of bags inside the bag house casing are used to filter the flue gas and remove the particles, including fly

101F26.EPS

Figure 26 Scrubber.

Figure 27 Bag house.

ash. Typically, power plants will have either a bag house or an ESP. Bag houses may include hoppers and a reverse air duct system used for cleaning the bags.

5.4.0 Selective Catalytic Reduction

Selective catalytic reduction (SCR) technology is used to reduce NO_x emissions. NO_x is an abbreviation for nitrogen oxides, including nitric oxide, nitrogen dioxide, and other nitrogen compounds produced by combustion. Nonselective SCR (NSCR) technology is also used to reduce NO_x emissions by injecting ammonia directly into the furnace. NO_x is contained in flue gases and is harmful to the environment. It causes acid rain and is a component of smog. An SCR injects a substance, such as ammonia or urea, into the hot flue gas. The resultant mixture then passes through a catalyst bed. In the bed, a chemical reaction turns the mixture into harmless nitrogen and water.

5.5.0 Water Waste

A power plant requires water purification and conditioning systems to prevent damage to plant equipment. Water is also used in the plant to generate and cool steam.

Cooling ponds are used to monitor or control the temperature effects of the cooling discharge from the condenser. Power plants along waterways are required to have discharge canals or ponds to ensure that the temperature of the water released back to the source complies with local, state, and federal regulations.

5.6.0 Combustion Waste Products

In addition to gases, light particles, and fly ash, coal combustion results in heavier solid waste. This waste drops to the bottom of the boiler and is collected for disposal. This waste is made up of ash and slag. See *Figure 28* for fuel waste collection points.

The ash is called hopper ash or bottom ash. It is collected in a water-filled hopper and flushed out of the collection unit, dried, and either sold to other industries or disposed of. Bottom ash is heavy, with a sandy texture. Only about 20 percent of the ash from coal combustion is this type of ash. The rest is fly ash. Bottom ash can be used as fill material for construction sites and roadways and is used as an aggregate in concrete and asphalt.

Slag is the molten part of the waste. It is also collected and cooled with water. The cooled slag becomes small, rough pellets. Because of its abrasiveness, slag is often used to make blasting grit.

Slag buildup on boiler tubes is a major problem in power plants. Slag formation is caused by the

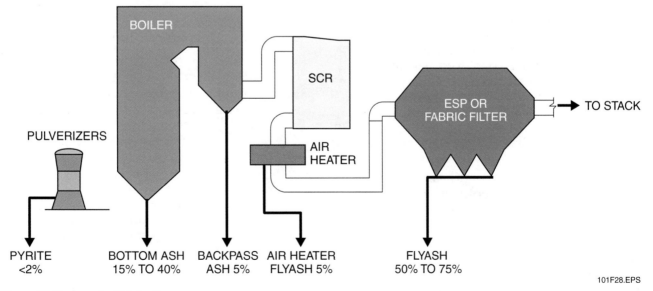

Figure 28 Fuel waste distribution.

buildup of molten ash that solidifies on the tubes and other surfaces. This reduces the efficiency of the boiler. Another problem with slag buildup is that large chunks of slag, referred to as clinkers, may break off and cause damage to internal boiler components as they fall.

Soot blowers and water lances are used to reduce slag buildup, but it still occurs. Deslagging to remove these deposits is commonly performed during a maintenance shutdown cycle. Removal methods include blasting with water under very high pressure, known as hydroblasting. Slag removal is often done by specialty contractors, some of whom use explosives to remove the slag.

6.0.0 ELECTRICAL TRANSMISSION AND DISTRIBUTION

Once the power plant generates electricity, it must be transferred to the consumer by a complex system of power lines. You see this system every day, everywhere you go. The power lines strung across the country are part of the electrical transmission and distribution system.

When electric power was first sold to customers, there was almost no transmission subsystem. Power plants were generally small and their customer base was confined to the local area. A single power plant supplied consumers with power, so outages were common. As technology developed, it allowed the length of transmission lines to increase, and electricity became an indispensible part of life. This created a demand for more power and better reliability.

As the industry grew, transmission lines of two or more power plants were connected. This provided consumers with more than one electrical source. This design improved the reliability of electrical power and helped utilities manage demand more easily. This was also the start of the modern power grid. Today, power is generated and transmitted to customers across a large and complex network that is commonly called the power grid.

The National Electric Reliability Corporation (NERC) was formed by act of Congress in 1969. Among the objectives was to create an electrical infrastructure that could prevent major power blackouts like the one that disabled the Northeastern U.S. in 1965. Over the years, the scope of its responsibility expanded to include parts of Canada and Mexico. The name was later changed to North American Electric Reliability Corporation. The function of the NERC is to develop and enforce reliability standards; monitor the bulk power system; assess adequacy annually via a 10-year forecast and winter and summer forecasts; audit owners, operators, and users for preparedness; and educate and train industry personnel.

In spite of common belief, the power grid in the United States is not a single continuous system of transmission and distribution lines. The power grid is divided into three main areas called the Eastern, Western, and Texas Interconnects (see *Figure 29*). These interconnects may be further divided into regional reliability councils and power coordinating commissions. Within any one of the three major interconnects, power is freely exchanged in order to meet changing demand. There are a small number of electrical connections among the three areas, but for the most part, they act independently. It takes a lot of equipment and many computers and people all working together to keep the power grid operating.

The transmission and distribution subsystems are made up of power lines, transformers, and substations. The transmission subsystem is responsible for transferring high-voltage AC from a power plant to the distribution subsystem. The transmission subsystem is further divided into transmission and subtransmission components, with the subtransmission placed between the transmission lines and distribution lines. The distribution subsystem distributes electrical power from the transmission system to the consumer.

The transmission and distribution of electricity from the power generating plant is a complicated and well-orchestrated system. The voltage level produced by the generator from the plant is increased at the power plant so it can travel effectively across the transmission lines. The voltage level is decreased when it reaches consumers. Typical voltage levels for each section are found in *Table 1*.

6.1.0 Transmission Subsystem

Transmission refers to the parts of the network that carry electrical power over long distances to a distribution substation. Power plants output varying voltages. A typical plant output is 13,800 volts. While this seems high, it is not enough to travel across miles of transmission lines without significant losses. To overcome voltage losses, the voltage is stepped up to an even higher voltage before it leaves the power plant.

The transmission and subtransmission components use transmission lines (*Figure 30*), transformers, switchyards, and substations to transmit energy to the distribution subsystem. In addition, substation operators monitor consumer energy use and route supplies as needed. This is

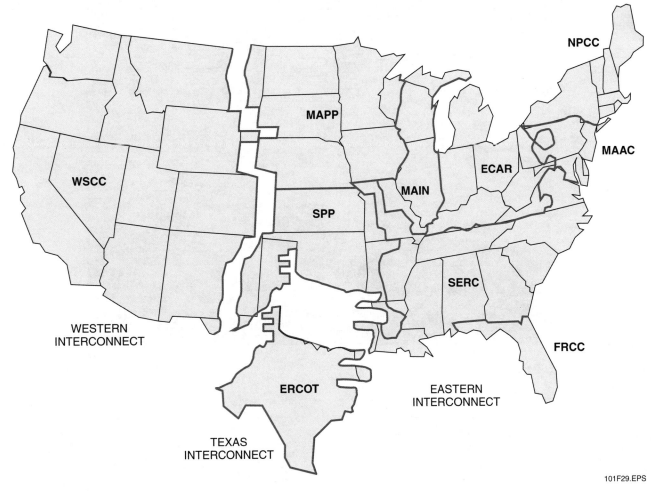

Figure 29 Power interconnect regions.

called load balancing. Further, the entire system is monitored for fault conditions. The distribution subsystem uses distribution substations and transformers to distribute power to customers.

The power grid uses step-up and step-down transformers to control voltage levels in various areas of the system. Step-up transformers increase the voltage level and step-down transformers decrease the voltage level. You will learn about how transformers work later in this course.

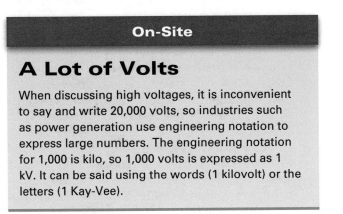

On-Site

A Lot of Volts

When discussing high voltages, it is inconvenient to say and write 20,000 volts, so industries such as power generation use engineering notation to express large numbers. The engineering notation for 1,000 is kilo, so 1,000 volts is expressed as 1 kV. It can be said using the words (1 kilovolt) or the letters (1 Kay-Vee).

Table 1 Typical Transmission, Subtransmission, and Distribution System Voltages

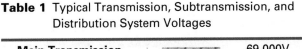

Main Transmission		69,000V
		138,000V
		220,000V
		345,000V
		500,000V
		750,000V
Subtransmission		13,800V
		23,000V
		34,500V
		69,000V
		138,000V
Distribution		2,400V
		4,160V
		13,800V
		23,000V
		34,500V

101F29.EPS

101F30.EPS

Figure 30 Transmission lines.

A substation can be located in a building, or it can be outdoors. Its purpose is to condition the voltage for further distribution, which may involve stepping up the voltage, stepping it down, or leaving it at the same level. Since the voltage at a substation is very high, the area is always fenced in to keep unauthorized people out of the danger zone.

6.1.1 Substations and Switchyards

Substations and switchyards are used to step up or step down voltage levels and to route electricity from one or more generating plants to various destinations. They consist of electrical buses, switching equipment, and transformers.

- *Power transmission switchyard* – Transformers located at the power-generation facility (*Figure 31*) are used to step up the generator output voltage for transmission to distant locations. The voltage produced by these substations can reach 765kV.
- *Subtransmission substations and switchyards* – These substations are located at transfer points in an electrical grid. They receive high-voltage power from a transmission substation or switchyard and step it down for distribution to local distribution substations. *Figure 32* shows a subtransmission substation where the voltage is stepped down from 230kV to 115kV using three single-phase transformers.

  A switchyard is a facility where power from different sources can be switched onto the grid and distributed to locations where it is needed.

101F31.EPS

Figure 31 Transmission substation transformer.

In the switchyard shown, there are two input lines in and two lines out. Either incoming line can be switched to either output line. Note the relay and control building at the lower left.

- *Distribution substation* – These substations (*Figure 33*) change the subtransmission voltage to the lower levels needed by end users. At this substation, 69kV is stepped down to 15kV.
- *Underground distribution substation* – Underground substations are also located near the end users. They perform the same function as aboveground substations. These substations are often located in urban areas where real estate is very expensive.

SUBTRANSMISSION SUBSTATION

69KV SWITCHYARD

101F32.EPS

Figure 32 Subtransmission substation and switchyard.

A grounding grid is a network of conductors placed in the soil beneath the substation. It is bonded to all metal noncurrent-carrying equipment and structures, such as equipment housings and fences. The purpose of the grounding grid is to safely route any high voltages that result from equipment switching, malfunctions, or lightning strikes.

The grounding and bonding systems in substations must carry electric current under both normal and fault conditions. A broken or disconnected connection is a potentially fatal hazard to a worker bridging the path to ground. It can also result in loss of service and equipment damage.

101F33.EPS

Figure 33 Distribution substation.

The grounding grid extends several feet outside the fence perimeter and is bonded to the fence at frequent intervals. In addition, the substation yard is covered with several inches of crushed gravel to provide a level of insulation or high resistance between a person and the soil. The aim is to get rid of stray voltages that can be hazardous to anyone approaching grounded parts during fault conditions.

6.2.0 Distribution System

The electrical distribution system is the final stage of electricity's journey to the consumer. Power in the range of 2.4 to 34.5kV is sent from a substation across distribution lines to a pole-mounted or underground transformer (*Figure 34*) which converts high-voltage electricity to a lower voltage that can be used at homes and businesses (*Figure 35*). Typical household consumers use 120/240 volts, while commercial and industrial consumers use 120/240/440 volts.

In urban areas, one transformer may serve as a distribution point for one or more neighborhoods, while in rural areas one transformer may serve only one home. Either way, the point where the electricity is used is called a load center.

In the past, the power transmission and distribution systems were aboveground, but today, some parts of the distribution system are underground. This adds a level of complexity to the system. Underground distribution systems are often found in congested urban areas where real estate is costly or in suburban housing developments where placing distribution lines underground makes the neighborhood more attractive.

Underground distribution systems have advantages and disadvantages. Underground lines are protected from weather, such as ice and wind from storms, but they are subjected to the possibility of flooding and corrosion. Underground systems take up no space aboveground, which is important in congested urban areas. However, installation of underground distribution lines is more costly than aboveground lines because of the trenching necessary to bury them. Further, since substations and transformers generate a great deal of heat, underground installations must be vented to the outside. Finally, damaged underground components are harder to find and repair, increasing outage periods.

PAD-MOUNTED TRANSFORMER FOR
UNDERGROUND SERVICE

POLE-MOUNTED TRANSFORMER

101F34.EPS

Figure 34 120V/240V power distribution transformers.

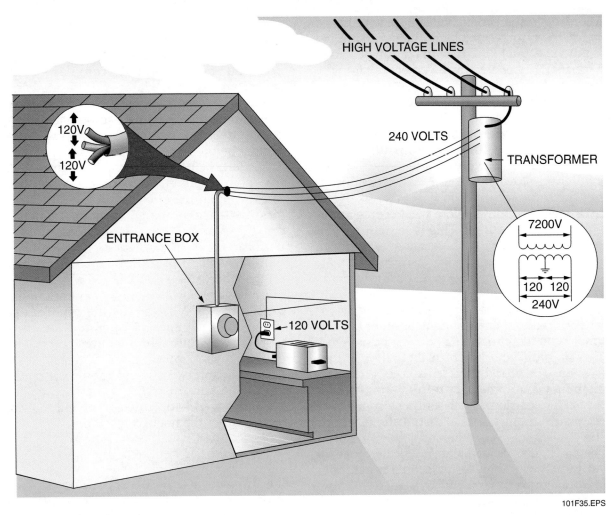

HIGH VOLTAGE LINES

120V
120V

240 VOLTS

TRANSFORMER

ENTRANCE BOX

120 VOLTS

7200V

120 120
240V

101F35.EPS

Figure 35 Residential power distribution.

7.0.0 CAREERS IN THE POWER INDUSTRY

Most workers in power generation and distribution jobs work for utility companies. However, many industries also require their own dedicated on-site power plants. These facilities can range from petroleum refineries, to hospitals, to large theme parks. These facilities need power-generation operators and maintenance personnel, as well as electrical distribution workers.

As with most professions, a certain amount of intelligence and training are needed to succeed in the power industry. However, as a power industry worker, you need to bring to the job characteristics that cannot be taught. They include the following qualities:

- *Responsibility* – Power industry workers routinely perform critical tasks that place themselves and others in danger. You need to take your job seriously.
- *Teamwork* – Potential hazards will surround you daily. You must be able to trust that your co-workers will look out for you, and they need to trust that you will be looking out for them.
- *Willingness to learn* – The power industry has changed enormously over its short life and will most likely change just as dramatically in the coming years. Power industry workers must be willing to learn about new equipment, procedures, and processes to keep up with technological advances.
- *Common sense* – This is not to be underestimated. Common sense tells you to step back when standing near open equipment or to stop and think before reaching out and touching something that could be energized. Common sense will help protect you until you gain experience.
- *Flexibility* – Be willing to adapt to changing needs. Employers want people who are willing to step up when special needs arise, which is a common occurrence in the energy industry. Employers need employees who are willing to work an odd shift, work overtime, or respond to a call-in request when an emergency comes up.

According to the United States Bureau of Labor Statistics, the number of jobs in the power industry is not expected to increase in the future. However, the majority of electrical workers fall into the age group of 35 to 54 years old. As these people retire, they will need to be replaced, so job prospects at all levels are excellent.

Although the concepts behind power generation and distribution are the same in all facilities, each work area will be very different. If you move to another employer, another facility, or another part of the same facility, you will most likely require extensive training before you become qualified in your new position. This is something you need to expect no matter how much experience you have.

Several tasks are common to all positions in the power industry. One is the need for continuing education. You will be required to attend training at your place of employment, another facility, or some type of school on a routine basis. During this training, you will learn about new energy policies and/or procedures, learn to operate and maintain new equipment, or refresh your present skills. Some positions, such as those in nuclear power plants, require special certifications.

One subject you will constantly be refreshing will be your safety knowledge and skills. Since all power industry employees work in a potentially dangerous environment, you will need to know and use safety procedures to perform your job. The details of these procedures will vary depending on the job and its location.

All employees, regardless of their position in the industry, will be required to take first aid training. Safety procedures in the power industry are so strict you may never need to use your training, but you and your co-workers should be ready if you ever need it.

It is important that each worker perform his or her job quickly and accurately to avoid serious damage to equipment, and unscheduled or prolonged outages. It is also important that maintenance and operations personnel work as a team.

7.1.0 Power Plant and Auxiliary Equipment Operators

Power plant operators control and monitor boilers, turbines, and generators. Auxiliary operators control and monitor equipment that is used to support the operation of the main power-generating components. These operators work together to spread power demands among generators. They also monitor operations to ensure that the plant produces enough electricity to meet the demand.

The plant supervisor manages the overall operation of the power plant.

These types of workers must operate and understand the functions of complex equipment, collect and analyze data, and initiate corrective actions as necessary. They also monitor the operation of power-generating equipment and must be able to determine when equipment is malfunctioning before a major failure occurs.

Modern plants usually have a centralized control room where plant operators work (*Figure 36*). These workers are often called control room operators. Auxiliary equipment operators work throughout the plant. They operate and monitor the performance of support equipment. Both positions require specialized training and a long period of on-the-job learning (OJL).

In nuclear power plants, operators must be licensed by the Nuclear Regulatory Commission (NRC). These types of operators usually start working as equipment operators or auxiliary operators. Nuclear power plant workers need formal education and experience. They must pass an examination administered by the NRC before they are authorized to operate equipment that controls the power of the nuclear reactor.

7.2.0 Maintenance

It is impractical to list all of the workers needed to support the operation of a single power plant, but they include welders, mechanics, boilermakers, electricians, electronics technicians, truck drivers, and heavy equipment operators, among others. Most of these workers require general training in their field and then some specialized training for their job at the power plant.

101F36.EPS

Figure 36 Power plant control console.

During power plant operation, systems in a power plant continually undergo routine maintenance where the operation of individual equipment components is monitored and adjusted. Equipment occasionally fails during operation. Maintenance workers must be able to quickly troubleshoot and repair the failed unit to allow plant operations to continue.

In addition, power plants must be shut down periodically for major maintenance or repairs. At this time, maintenance workers will perform tasks that cannot be done while the plant is operating. Often, temporary workers will be brought in to perform work and the plant's permanent staff will be placed in supervisory positions to ensure that the work is done quickly and safely.

7.3.0 Fuel Handling

Fuel handling (*Figure 37*) is an important part of power plant operations. Fuel handling includes storage, transportation, and delivery of fuel. Fuel handlers operate and maintain fuel-handling and dust-suppression equipment. Since the moisture content of coal is one of the most important factors in achieving a good burn, handlers will constantly monitor coal storage conditions. They test coal before it is transported to the power plant to ensure that it does not contain too much moisture.

7.4.0 Laboratory Analyst

A key element in the operation of a power plant is the chemical analysis of boiler water and flue emissions. Specialized workers called analysts perform these tests. Based on the results of the tests, analysts may order adjustments to water conditioners, as well as emissions and filtration equipment. These workers require general train-

ing in the field and further training for their job at the power plant.

7.5.0 Line Workers

Line workers can include tree trimmers, pole erectors, and power line technicians. They install, maintain, and repair the power lines that connect generating plants to consumers. This can include everything from erecting power poles (*Figure 38*) and digging trenches to repairing damaged power lines (*Figure 39*). Power line work is very physical and is performed outdoors in all sorts of weather.

Line workers must be able to operate a variety of equipment, such as trenchers, augers, and cranes, as well as hand and power tools. They must also operate a wide variety of test equipment to verify that an installation is correct and to troubleshoot faults. Line workers cannot be afraid of heights because much of their work is performed high above ground. Some line worker jobs require a great deal of physical strength.

Line workers require special training on transformers, electrical power distribution systems, and substations. All line workers must be trained to understand and follow safety procedures, because they are constantly exposed to serious hazards on the job.

Line workers require several years of on-the-job learning and classroom work to become proficient. Formal apprenticeships are common. Employers look for people with basic knowledge of algebra and trigonometry, as well as good reading and writing skills.

101F37EPS

Figure 37 Coal-handling system.

101F38.EPS

Figure 38 Erecting a power pole.

Figure 39 Power line worker.

101F39.EPS

7.6.0 Energy Brokers

Energy brokers buy and sell bulk electricity. They may work for power-generating plants, utilities, corporations, private marketing agencies, or federal and state governments. Each employer can have different requirements for education and experience. Brokers are licensed by the states in which they operate.

8.0.0 SAFETY REGULATIONS

Working in the electrical power-generating industry can be very dangerous, but the industry has an excellent safety record. This is because the industry constantly monitors, reviews, and enforces safety policies and procedures. It ensures that industry workers are properly trained to safely perform their jobs and follow all safety procedures.

Safety is important in the power industry and it goes far beyond the classroom. Safety is everyone's responsibility. During your career, you will constantly attend safety briefings, follow procedures that will include safety precautions, use personal safety equipment, and train others to be safe. These measures are in place to protect you and your co-workers from injury or death and to prevent damage or destruction of equipment.

You will find that your employers are very serious about safety. Most sites will have policies and procedures which specifically state that departure from the policy is grounds for disciplinary action. This could mean termination from your job.

It is very important that you know and follow all safety policies and procedures that regulate the performance of your job. These policies and procedures will keep you safe while performing your normal duties, as well as when you find yourself in unfamiliar situations.

8.1.0 Government Agencies Involved With Safety

In addition to your employer's requirements and industrywide safety requirements, federal and state governments regulate safety in power-generation plants and transmission and distribution systems. Some of these agencies include the following:

- Occupational Safety and Health Administration (OSHA)
- Mine Safety and Health Administration (MSHA)
- American National Standards Institute (ANSI)
- Transportation Security Administration (TSA)
- Environmental Protection Agency (EPA)
- Department of Transportation (DOT)
- Nuclear Regulatory Commission (NRC)

8.1.1 Occupational Safety and Health Administration

The Occupational Safety and Health Administration (OSHA) was created in 1970 to provide education and guidelines to employers and employees to prevent work-related injuries, illnesses, and deaths. The guideline that covers electric power generation, transmission, and distribution is *OSHA Standard 1910.269*.

All hazardous workplaces with 11 or more employees fall under OSHA guidelines. These workplaces must provide employees with information and training about how to stay safe on the job. In addition, the workplace must make it possible for workers to stay safe. That means the facility and its equipment must be kept in good working order.

Your employer has spent a great deal of time and money developing and enforcing a safety program. These procedures are written by workers who know about the operation of equipment, how the equipment fits into a system, and how the system fits into the facility. Every site will have different ways to handle the writing, approval, and performance of a safety procedure.

Your job is to follow procedures. It is up to you to follow each procedure as written every time you perform the task. It is fine to recommend a change to a procedure. It is also okay to recommend a modification to a safety device used during a procedure. But it is never acceptable to omit a procedure, modify a procedure on your own, or bypass any safety device.

8.1.2 Mine Safety and Health

The Mine Safety and Health Administration (MSHA) oversees all mining and mineral processing operations in the United States, regardless of size, number of employees, or method of mining. Because mining operations often have an on-site power plant, you may be employed at a facility that falls under the jurisdiction of MSHA.

MSHA develops and enforces safety and health rules at all U.S. mines by providing technical and educational assistance. MSHA works with industry, labor, and other federal and state agencies to improve safety and working conditions for all mineworkers.

MSHA conducts mine inspections, accident investigations, and training. It coordinates the development and revision of safety and health rules as required, and it imposes penalties against mine operators for failing to follow health or safety requirements.

8.1.3 American National Standards Institute

The American National Standards Institute (ANSI) oversees the voluntary standards system in the United States. Its original purpose was to establish standardization of similar components. Before ANSI was formed, factories manufactured components to their own specifications. Parts from different suppliers were not compatible. The first task ANSI undertook was establishing guidelines to standardize pipe threads. From there, ANSI developed and established safety standards for home and commercial equipment.

ANSI publishes the *National Electrical Safety Code, IEEE/ANSI C2*, which the power generation industry follows to safeguard workers during the installation, operation, and maintenance of electric supply and associated equipment. This code was developed by the Institute of Electrical and Electronics Engineers (IEEE). ANSI has worked with the National Fire Protection Association (NFPA) to produce several documents that specify ways to prevent electrical fires and avoid electrical hazards.

8.1.4 Environmental Protection Agency

The EPA is tasked with developing and enforcing policies that protect human health and safeguard air, water, and land. The EPA makes its own policies and has input into the policies of other agencies concerning natural resources and human health. These include policies for the power generation industry.

EPA policies that affect the power industry include those covering flue gas emissions, wastewater treatment and disposal, and radioactive waste disposal. In addition, the EPA tracks emissions from power plants and the types of energy sources used in power plants by state and region.

8.1.5 Department of Transportation

The United States Department of Transportation (DOT) is tasked with ensuring a fast, safe, efficient, accessible, and convenient transportation system that serves our national interests and improves the quality of life of its users. The DOT evaluates current transportation needs as well as future needs. The federal government has a DOT and so does each state. They work together to ensure the safety of our transportation system.

When you think of the DOT, you most likely think of the Federal Highway Administration or Federal Aviation Administration. While these two agencies are important parts of the DOT, it is actually divided into ten administrations, which also include the following:

- Federal Railroad Administration
- Maritime Administration
- Pipeline and Hazardous Materials Safety Administration

These administrations oversee the safe transportation methods used to carry energy sources to and waste products from power plants. Plant personnel responsible for these tasks will need to work with the DOT to ensure the safety of all systems and processes.

Because the DOT is responsible for a safe transportation system, it has the authority to inspect those parts of power plants that fall under transportation. This is chiefly fuel pipelines, but also includes wastewater pipelines, hazardous waste removal, and uranium transportation. The DOT sets forth policies and procedures for these systems and processes. They include maintenance, operation, and inspection standards. In addition, the DOT has the authority to issue citations for non-compliance that could lead to fines and unscheduled shutdowns.

8.1.6 Nuclear Regulatory Commission

The U.S. Nuclear Regulatory Commission (NRC) was created to allow the safe use of radioactive materials while ensuring that people and the environment are protected. The NRC governs the operation of commercial nuclear power plants through licensing, inspection, and enforcement of its regulations. The NRC is responsible for the following:

* Reactors used for generating electric power, as well as those used for research, testing, and training
* The use of nuclear material in industrial and academic settings, and facilities that produce nuclear fuel
* Transportation, storage, and disposal of nuclear materials and waste
* Decommissioning of any nuclear facilities

The NRC sets policies and procedures for all aspects of nuclear power generation, including licensing of new plants; renewing the licenses of existing plants; licensing of sites that manufacture nuclear materials; and management of all nuclear waste. In addition, the NRC tests and certifies nuclear power plant operators and issues security clearances to workers as needed. The NRC has the authority to inspect nuclear facilities, review plant procedures, issue fines for non-compliance, observe you doing your job, revoke a worker's certification, and shut down a facility (temporarily or permanently).

8.1.7 Transportation Security Administration

The Transportation Security Administration (TSA) was formed after the September 11 terrorist attacks in 2001. It is part of the Department of Homeland Security. The TSA is tasked with the protection of the nation's transportation systems. One of its programs is the Transportation Worker Identification Credential (TWIC), which is a security clearance program designed to prevent unauthorized entry into the nation's maritime transportation system.

This program affects workers at power plants that receive coal by water. Each site will have different policies for TWIC, depending on how the facility is set up. Some sites will require all workers to obtain this clearance, while others will require only those who have access to the port to have it.

To obtain the TWIC clearance you will be required to provide personal information (name, address, Social Security number, etc.) and positive identification; sign a disclosure form that authorizes a criminal background check; provide a complete set of fingerprints; have a digital photograph taken; and pay a fee. In some cases your employer will pay the fee for you, but most likely a TWIC clearance will be a condition of your employment, so you will need to pay the fee. The clearance is valid for five years and is transferrable between jobs.

9.0.0 SAFETY PRACTICES

Workers in all industries can suffer from accidents. Anyone can slip and fall, cut a finger or hand, or bump into a door. Power plants have additional hazards. The following are some of the unique hazards in the power industry:

* *Electrical* – Electrocution or electric shock from power lines, generated power, stored power, arcing, and malfunctions.
* *Pressure* – Explosions due to steam or air buildup, fuel vapors, chemicals, or dust.
* *Chemical* – Inhalation injuries or death from chemicals, solvents, or coal dust.
* *Mechanical* – Crushing injuries from pulverizers and heavy equipment.
* *Thermal* – Burns from steam, fuel, hot ash or coal, chemicals, or electrical sources.
* *Combination* – Some hazards set off a chain reaction that exposes workers to multiple dangers. A transformer explosion from overheating can cause burns, injuries from flying debris, an electrical fire, and/or electrocution.

9.1.0 Policies and Procedures

Power plant operations are not left to chance. Each facility will have specific, written procedures for the performance of routine repetitive tasks such as daily equipment operation checks and adjustments. They will also have general policies covering unique situations. For example, when you are troubleshooting malfunctions, you may not have a detailed procedure for each system in the plant, but you will have a general policy or procedure that is designed to keep you, your co-workers, and the equipment safe. You will often need to follow several procedures to perform a single task. Each procedure will specify what is to be done and who is authorized to order the performance of each task. Because of electrical hazards, each plant will have specific procedures that cover isolation, lockout/tagout, and clearances.

It is possible to keep yourself safe, even in unfamiliar situations, by answering the following questions before you take any action on the job:

* Have you been trained to use this device or equipment?

- Are you authorized to perform the task?
- Do you know why you are performing a procedure?
- Do you understand the purpose of the action?
- Do you know what to do if there is a malfunction because of your action?

If you do not answer "yes" to each of these questions, stop what you are doing and ask your supervisor or an experienced co-worker for help. When it comes to safety, you only get one chance.

9.2.0 Foreign Material Exclusion

Anyone working in a power plant must comply with foreign material exclusion (FME) policies. The purpose of an FME program is to prevent damage to equipment from stray objects. Foreign material includes such items as coins, eyeglasses, tools, rags, oil, hardware, vegetative matter, fish, stones, and ice. The FME program is applied to any area where critical plant systems and components are located.

When work is performed near critical components, an FME area will be set up, a job plan will be made, and all involved employees will be briefed before any work is started. The job plan outlines the performance of the job, including the use of any tools, the removal of any parts, and the generation of any waste such as shavings, used lubricants, or anti-static packing material. Based on the job plan, workers will assemble tools, equipment, and other items necessary to perform the task.

The FME area will be designated and its boundaries will be clearly marked with signs and barriers (*Figure 40*), usually a distinctive colored tape. Entry and exit points will be clearly marked. At each of these points, logs will be placed to record material entering and leaving the FME area. The purpose of these policies and procedures is to prevent stray items from being left in or near a critical system or component where they can cause damage.

All workers are responsible for FME, even when they do not actually enter the exclusion area. For example, if you help a co-worker gather items to be carried into the area, you are subject to the FME regulations. An item as harmless as a scrap of plastic could stick to a metal chassis. Those entering the exclusion area must be especially aware to prevent foreign materials from entering the area. The following rules must be observed:

- Inspect all items for loose parts, defects, and contamination such as oil film, moisture, and dirt particles. Replace suspect items and clean contaminated items before entering the FME area.

- Remove all unnecessary packing materials and paperwork from replacement parts, including loose connector covers.
- Inspect all equipment to be certain that all hardware is secured and no foreign material is clinging to the chassis.
- Remove any unnecessary items from your person, including earrings, coins, watches, and keys.
- Inventory all items going into the area, including pens, pencils, and eyeglasses.
- Use taglines and lanyards whenever possible on items entering the FME area (see *Figure 41*).
- Carefully inspect the area for foreign matter upon entering and before leaving the area.

FME BARRIER TAPE

FME BARRIERS

101F40.EPS

Figure 40 FME barrier.

- Before you begin work, cover any openings that could collect foreign material (see *Figure 42*).
- Before leaving the area, flush or blow out pipes and other units as needed.
- Inventory all items, including waste, as they leave the area.

When the work is completed and before the equipment is cleared for operation, the inventory logs and job plan must be reviewed to be sure that all materials, including waste, were removed from the area.

9.3.0 Grounding

All electrical equipment must be properly grounded to the grounding grid at the facility when it is installed. However, the straps and cables connecting the equipment to the station's grounding grid can become damaged or loosened over time. Good maintenance programs include routine inspection and testing of grounding systems to ensure that the facility equipment remains properly grounded.

A ground fault is a situation in which electrical energy from any source finds its way through an unintended path to ground. There are a number of other definitions of a ground fault, but the bottom line is that if you become the unintended path, you can be injured, or even killed. Part of the reason for grounding de-energized circuits and equipment is so that work can be done on them safely.

Electrical workers must often work on equipment that they have temporarily de-energized. Some of this equipment may be located near energized equipment that could induce voltage into the de-energized equipment. Temporary grounding devices are installed on the de-energized equipment to protect workers from electric shock. Electrical energy will follow any path to ground. The proper name for these devices is temporary protective grounds (TPGs).

OSHA guidelines require that employers provide their workers with a safe work environment. When electrical equipment is involved, the equipment and all inputs/outputs must be placed into an electrically safe work condition. This means that all equipment must be de-energized, any stored electrical energy must be discharged, and the equipment must be locked out before any work can be performed. After the electrical equipment has been de-energized, TPGs are installed to provide grounding paths to help protect anyone working on the equipment. Temporary protective grounds are also called personal protective grounding (PPG) because that is their primary purpose. They provide an alternate path to ground so that any unwanted electrical energy will follow that alternate path. *Figure 43* shows a worker installing a TPG at a substation and a power line worker protected by TPGs on the power lines.

ADJUSTABLE LENGTH WITH CAM LOCK

PLASTIC CLIP

QUICK DISCONNECT

101F41.EPS

Figure 41 Hard hat lanyard.

101F42.EPS

Figure 42 FME cover.

Figure 43 Examples of temporary grounding.

101F43.EPS

10.0.0 POWER INDUSTRY ECONOMICS

Economics is the study of the production, distribution, and consumption of goods and services. Electricity is a product that can be bought and sold. Electricity is a vital part of our lives, so it is critical that the power supply be reliable and affordable. In this section, you will read about the cost of generating and distributing electricity, ways in which the industry ensures that electricity is reliable, and some of the government agencies that make policies to ensure that electrical power is reliable.

Electric power plants generate electricity. Utility companies generally transmit and distribute electricity. Both can be owned by taxpayers, investors, a cooperative, or by the federal government. Taxpayer-owned facilities are publicly owned facilities that were built with taxpayer funds. Investor-owned utilities are owned by investors, who may be private citizens, municipalities, or corporations. Investor-owned utilities are based on shares that can be bought and sold. Cooperatively owned utilities are owned by the customers of the utility. Federally owned utilities are owned by the United States government. All utilities are regulated by local, state, and federal authorities, and in the case of many electric cooperatives, by their boards of directors.

10.1.0 Laws

To ensure the integrity of the entire electrical system, federal and state governments regulate the operation of the system. Before 1992, electricity was supplied to consumers by regional generating facilities using transmission and distribution lines owned by the utility. Consumers had no real choice of electrical generation facility. They simply used the utility in their area.

In 1992, Congress passed laws to partially restructure the power industry in an effort to promote competition between electricity suppliers and to reduce energy costs to consumers. The Energy Policy Act (EPACT) of 1992 permits any electrical power plant to use a utility's transmission and distribution lines. This act separated the generation of electricity from its transmission and distribution. This regulates transmission and distribution of electricity while deregulating generation of electricity. The purpose of EPACT was to increase competition among electricity generators and decrease costs to consumers.

Regional utilities still maintain and support the transmission and distribution lines, but they do not necessarily generate the power going across them.

10.1.1 Federal Energy Regulatory Commission

The Federal Energy Regulatory Commission (FERC) regulates the electrical transmission system, approves sale or leasing of transmission facilities, and approves mergers and acquisitions between independently owned utilities. FERC regulates electrical power transactions between states. These types of transactions are called interstate transactions. Interstate transactions are sometimes very complex because they involve laws in both states. In addition, FERC licenses hydroelectric facilities on navigable waterways.

FERC has jurisdiction over electric power policies that affect the reliability of electrical service to customers in the continental United States, as well as parts of Canada and Mexico. As a federal agency, FERC has jurisdiction over the NERC, which was established by the power industry. The NERC is divided into ten regions called councils.

The councils are responsible for overall coordination of bulk power policies that affect the reliability and adequacy of service in their areas. They also regularly exchange operating and planning information among their member utilities. The boundaries of the NERC regions follow the service areas of the electric utilities in the region. Many of the regions do not follow state boundaries.

10.2.0 Costs

Before deregulation, costs related to the generation of electricity were passed on to the consumer. These costs included the construction of new facilities and the operation and maintenance of existing facilities. Naturally, consumers could not pay all of these costs as they were incurred, so they were recovered over time in the electricity rates charged to consumers.

Building new power plants and transmission and distribution facilities is expensive. This is a problem for deregulated power plants because to be competitive, the rates must be as low as possible. New power plants have construction costs to recoup, as well as operating costs, so pricing rates low enough to be competitive is difficult. This is a very real problem in the power industry. The term for not being able to recover a generating plant's construction costs is stranded costs.

Construction costs are one of the reasons that alternatives to fossil fuel are not built rapidly.

Construction is financed either by private investors or by taxes. Private investors want a return on their investments, while taxpayers often resist paying more taxes.

Generating electricity is costly, too. Once facilities are running, the greatest costs are for maintenance and operation. Maintenance keeps the equipment and the facility operating reliably, safely, and efficiently. Operation includes all of the tasks that are needed to maintain and monitor plant facilities.

One of the greatest operating costs of electrical power generation is the cost of fuel. It is important for power plants to have a ready and plentiful fuel source. Rolled into fuel prices is the cost of transporting fuel to the generating facility.

Coal is typically moved by trains and barges, which use diesel fuel for power. Diesel fuel is a petroleum product, so when oil prices are high, coal prices are high. Natural gas and oil are transported to power plants through pipelines. A long pipeline is expensive to build and maintain. For this reason, generating facilities are usually built near their source.

10.3.0 Reliability

Most of the commercially available electricity used in the United States is generated as it is used. Smaller power-generating facilities, such as those using solar and wind energy, are able to store power and release it on demand. But there are very few of these, and at this time they are not commercially practical.

The amount of power a plant can produce is measured in megawatts. One megawatt is equal to one million watts, and can typically supply power to about 1,000 homes, depending on location and power use. The Chief Joseph plant in Washington is hydropowered and can produce more than 2,500 megawatts of power. Plant Scherer in Georgia is coal-powered and can produce more than 3,500 megawatts of power. These are among the largest plants of their type; other plants of the same type may output only one or two megawatts. A typical large hydroelectric plant produces 135 megawatts of power.

Electrical power generation and distribution is very complex. Think about the use of electricity for a moment. During the day, most people are awake and at work. Schools are in session, factories are in full operation, retail stores and offices are open and doing business. At those times, there is a high demand for electricity. On hot summer afternoons, air conditioners are running almost constantly, increasing electrical power demands. At night, the electrical demand usually decreases.

However, there is always a demand for electricity—even at night. This is called the base load. As people use electricity, its demand increases. This is called the intermediate load. As demand peaks, more electricity is required. This is called the peak load. The operators of the power grid must be able to supply all of these energy needs efficiently.

Unfortunately, power plants cannot supply unlimited quantities of power at any moment. Each power plant has a maximum amount of energy it can supply. This is called full power. It takes time to bring a power plant up to full power, and this time varies between plants. In addition, the cost to run each plant varies.

Nuclear power plants are stable at full power, so they are usually always operating (online). Nuclear plants normally support base load requirements. Other power-generating facilities are brought on line based on demand and economics.

Combustion turbines handle changing loads well, so they are used as intermediate plants. Plants that are older, smaller, and most costly to run (usually coal- or oil-burning plants) are used for peak hours. These plants are often called peakers.

The electric power system in the United States is maintained and monitored by dispatch centers. Demand for electricity changes constantly. Operators at these centers need to ensure that demands are met by buying and selling electricity as needed. In addition, operators need to maintain the integrity of the distribution system by ensuring that the power-handling capability of the transmission and distribution lines is not exceeded.

The importance of continuous electrical power is such that great measures are taken to avoid outages. Consider the impact of a blackout or brownout.

A blackout is a total power loss. It can cause refrigerators in homes and businesses to stop working, causing food to spoil. This can represent a sizable financial loss, especially to businesses. It can cause lights in buildings to go out and elevators to stop working, causing safety issues. Blackouts can cause computers and other electronics to shut down, preventing workers from performing their duties. This can result in lost work, as well as lost wages.

A brownout is a decrease in system voltage. While brownouts are not as serious as blackouts, they can place stress on electronic equipment and motors. Blackouts and brownouts can be caused by equipment failures, accidents, normal wear, natural disasters, or too great a demand for electricity.

Some outages are caused by mechanical failures. Most of these types of failures can be avoided by regularly inspecting all equipment and replacing or repairing defective parts. Part of your job will most likely be to perform routine maintenance. The purpose of this type of maintenance is to detect faults and ensure that all equipment, including power poles, is operating at peak efficiency.

Some outages are caused by nature. Ice storms and strong winds can down power lines above the ground, and heavy rains can flood lines below ground. Energized equipment can be severely damaged by short circuits and flooding, so power providers in coastal areas expecting severe weather from a hurricane will often shut down a facility before the severe weather arrives. Even this type of scheduled outage—though necessary—is only done after much investigation and discussion.

Rolling blackouts and brownouts are control methods used to take the stress off an electrical system due to high demand for electrical power. These are emergency measures taken by electrical power providers to protect equipment. They may occur with or without warning and are used only as a last resort to avoid a total blackout. They are often scheduled for peak demand periods on very hot days.

11.0.0 The Future of the Power Industry

Every decade has brought changes to the power industry and that is likely to continue throughout your career. Environmental considerations, improved technology, and fuel availability will most likely drive future changes in the power industry.

11.1.0 Environmental Considerations

As technology advances, the power industry will use renewable resources to generate electricity on a commercial basis. Federal and state governments are now providing funding in the form of grants, loans, and incentives for the development of environmentally friendly energy alternatives.

As a worker in the power industry, you will need to continually update your skills to stay current with technology. In the past, industry advancement was centered largely on fuel sources. This will likely continue as technology to use wind, solar, and other renewable energy resources becomes cheaper and more reliable.

Some advances will involve new and better ways to capture, recycle, and otherwise reduce harmful byproducts from renewable and non-renewable fuel sources.

No fuel source is perfect. Widespread use of any source will have at least some undesired effects. Biomass fuels use renewable sources, but produce carbon dioxide. Wind-powered turbines can kill whole flocks of migrating birds. Solar devices contain components that can be harmful to the environment.

11.2.0 Future Technology

At this point, it is impossible to determine which path the energy industry will take. It is important to expect the unexpected. Every year brings new and exciting technologies that have the potential to dramatically change the industry. Some of these technologies will not work out and others will.

In the late 1800s and early 1900s, Nikola Tesla investigated ways to transmit electricity without using transmission lines. He had limited success. At that time, the technology was not developed, but in 2007, researchers at the Massachusetts Institute of Technology were able to light a 60W bulb located 6.5 feet (2 meters) from the energy source using wireless technology.

Another technology currently under investigation is the use of hydrogen as an energy source for fuel cells. Fuel cells are similar to batteries in that they use a chemical reaction to generate electricity. Batteries are self-contained units that have all the chemicals needed to generate power, so when the chemicals are used up, the battery is dead. Hydrogen fuel cells get their energy from external sources.

Hydrogen fuel cells use hydrogen and oxygen to generate electricity and the only waste product is water. In addition, the efficiency of a hydrogen fuel cell can approach 70 percent. The disadvantages are that the technology is expensive and that its fuel, hydrogen, is typically a byproduct of natural gas processing. Further, hydrogen is highly flammable just like gasoline. Developmental research is exploring cost-effective methods that use hydrogen from the air.

Coal gasification is called clean coal technology. It breaks down coal into its chemical components. The process involves exposing coal to steam and carefully controlled amounts of oxygen under high temperatures and pressures, causing chemical bonds to break and form new ones that produce gases, like carbon monoxide, hydrogen, and other gaseous compounds. The gases can be used to power gas and steam turbines in combined cycle plants. The hydrogen gas can be used to power fuel cells. The advantage of this technology is that it removes most of the sulfur and nitrogen from the coal, producing a clean gas that is comparable to natural gas. Coal gasification electric power plants are now operating commercially in the United States.

One future consideration is to decrease the demand on our existing power grid by using smart-grid technology. The smart grid uses two-way digital communications to monitor power availability and electrical flow through the grid. Then, based on power availability, it controls the operation of electrical devices in homes and businesses. This system turns on appliances when power is abundant and turns them off when power is not. Using a system such as this will decrease the demand on the grid and power producers, which in turn will reduce cost and increase reliability.

As part of the smart grid, there is some research into using superconductive transmission lines. Superconductivity is when resistance drops to zero at low temperatures. Superconductive transmission lines reduce power loss.

12.0.0 STANDARDIZED TRAINING BY NCCER

The National Center for Construction Education and Research (NCCER) is a not-for-profit education foundation established by the nation's leading construction companies. NCCER was created to provide the industry with standardized construction education materials, the Contren® Learning Series, and a system for tracking and recognizing students' training accomplishments—NCCER's National Registry. Refer to the *Appendix* for examples of NCCER credentials.

NCCER also offers accreditation, instructor certification, and skills assessments. NCCER is committed to developing and maintaining a training process that is internationally recognized, standardized, portable, and competency-based.

Working in partnership with industry and academia, NCCER has developed a system for program accreditation that is similar to those found in institutions of higher learning. NCCER's accreditation process ensures that students receive quality training based on uniform standards and criteria. These standards are outlined in NCCER's Accreditation Guidelines and must be adhered to by NCCER Accredited Training Sponsors.

More than 450 training and assessment centers across the U.S. and eight other countries are proud to be NCCER Accredited Training Sponsors. Millions of craft professionals and construction managers have received quality construction education through NCCER's network of Accredited Training Sponsors and the thousands of Training Units associated with the Sponsors. Every year the number of NCCER Accredited Training Sponsors increases significantly.

A craft instructor is a journeyman craft professional or career and technical educator trained and certified to teach NCCER's Contren® Learning Series. This network of certified instructors ensures that NCCER training programs will meet the standards of instruction set by the industry. At the time of this writing, there are more than 4,300 master trainers and 45,000 craft instructors within the NCCER instructor network. More information is available at www.nccer.org.

12.1.0 Apprenticeship Training

Formal apprenticeship programs conform to federal and state requirements under *CFR Titles 29:29 and 29:30*. All approved apprenticeship programs provide OJL as well as classroom instruction. The related training requirement is fulfilled by all NCCER craft training programs. The main difference between NCCER training and registered apprenticeship programs is that apprenticeship has specific time limits in which the training must be completed. Apprenticeship standards set guidelines for recruiting and outreach, and a specific time limit for each of a variety of OJL tasks. Additionally, there are reporting requirements and audits to ensure adherence to the apprenticeship standards. Companies and employer associations register their individual apprenticeship programs with the Office of Apprenticeship within the U.S. Department of Labor, and in some instances, with state apprenticeship councils (SAC). OJL of 2,000 hours per year and a minimum of 144 hours of classroom-related training are required.

The traditional apprenticeship program required OJL of 2,000 hours per year and a minimum of 144 hours of classroom-related training. Apprenticeship programs therefore varied in length from 2,000 hours to 10,000 hours. However, on October 29, 2008, the U.S. Department of Labor published new regulations to modernize the National Apprenticeship System. These regulations provide for more flexibility in how related technical instruction (RTI) can be delivered. They also provide registration agencies with the option of issuing interim credentials to offer active apprentices official recognition of their accomplishments and equip them with a portfolio of skills and incentives to complete their programs and continue their career preparation. Finally, these regulations allow program sponsors to offer three different ways for apprentices to complete a registered apprenticeship program:

- The traditional, time-based approach, which requires the apprentice to complete a specific number of OJL and RTI hours.
- A competency-based approach, which requires the apprentice to demonstrate competency in the defined subject areas and requires OJL and RTI.
- A hybrid approach, which requires the apprentice to complete a minimum number of OJL and RTI hours and demonstrate competency in the defined subject areas.

12.2.0 Youth Training and Apprenticeship Programs

Youth apprenticeship programs are available that allow students to begin their apprenticeship or craft training while still in high school. A student entering the program in the 11th grade may complete as much as one year of the NCCER training program by high school graduation. In addition, programs (in cooperation with local construction industry employers) allow students to work in the craft and earn money while still in school. Upon graduation, students can enter the industry at a higher level and with more pay than someone just starting in a training program.

Students participating in the NCCER or youth apprenticeship training are recognized through official transcripts and can enter the second level or year of the program wherever it is offered. They may also have the option of applying credits at two-year or four-year colleges that offer degree or certificate programs in their selected field of study.

SUMMARY

Although the potential value of electricity was known for hundreds of years, it wasn't until the 20th century that scientists and engineers were able to make electricity available for practical use on a large scale. Today, electricity is an essential element in the lives of everyone in the developed world.

The majority of electrical power in the U.S. is generated by fossil-fuel power plants that burn coal, oil, or natural gas. Of these fuels, coal is the most common. It is recognized by both the government and the power industry that we will eventually run out of fossil fuels. Nuclear and hydroelectric power are part of the solution, but they are not the total solution. Alternative, cleaner forms of energy, such as solar and wind power, are being brought into more common use. Today, these sources are providing power to meet peak demands. In the future, they will play a much larger role in meeting the nation's power needs.

A significant advancement in the second half of the 20th century was the creation of a national power grid. This grid allows power to be managed at a regional level. This means that power can be bought by areas with a high demand and sold by those who have surplus capacity.

The power industry requires tens of thousands of workers to operate and maintain power plants and to install, maintain, and repair power transmission and distribution equipment and facilities. Because of the unique hazards involved, safety is a critical issue in the power industry. Several government agencies regulate safety in the power industry. In addition, every employer in the industry has safety practices that their employees must follow.

Working in the power industry requires extensive knowledge that can only be acquired through training. The training you receive in this program will help prepare you for a career that is both demanding and rewarding.

Review Questions

1. Losses of AC voltage across long transmission lines are minimized by using _____.
   a. transformers
   b. converters
   c. isolators
   d. amplifiers

2. Forms of energy include _____.
   a. electrical, chemical, thermal, mechanical
   b. electrical, water, thermal, mechanical
   c. electrical, chemical, coal, mechanical
   d. electrical, chemical, gas, mechanical

3. Potential energy and kinetic energy are the same thing.
   a. True
   b. False

4. A nonrenewable energy source is _____.
   a. geothermal
   b. hydrogen
   c. natural gas
   d. hydropower

5. In the United States, most power-generating plants are powered by _____.
   a. wind
   b. water
   c. natural gas
   d. coal

6. In a coal-fired power plant, the power to run the generator is developed by _____.
   a. high-pressure steam applied directly to the rotor of the generator
   b. high-pressure steam applied to the rotor of a steam turbine
   c. high-pressure steam applied to a combustion turbine
   d. the conversion of coal to natural gas

7. A condenser is used to _____.
   a. supply air to the burner
   b. turn steam into water
   c. supply air to the gas jet
   d. initiate nuclear fission

8. A combustion turbine uses steam produced by a gas-fired boiler.
   a. True
   b. False

9. The common fuel source for nuclear power plants in the U.S. is _____.
   a. uranium
   b. copper
   c. potassium
   d. clinker

10. Ash and other solid particles in flue emissions can be captured with a(n) _____.
   a. electrostatic precipitator
   b. scrubber
   c. pulverizer
   d. gasification unit

11. Flue emissions pass through scrubbers to remove _____.
   a. carbon dioxide
   b. carbon monoxide
   c. sulfur dioxide
   d. sulfur monoxide

12. Flue gas is channeled through a selective catalytic reduction (SCR) device to remove _____.
   a. carbon dioxide
   b. sulfur dioxide
   c. nitrogen dioxide
   d. hydrogen dioxide

13. Typical voltages distributed to consumers are _____.
   a. 69 and 120 volts
   b. 69 and 240 volts
   c. 120 and 240 volts
   d. 138 and 440 volts

14. Installing power distribution lines underground _____.
   a. protects them from damaging storms
   b. makes repairs easier than above-ground installations
   c. is cheaper than installing them above-ground
   d. decreases troubleshooting and repair time

15. Underground substations must be _____.

    a. vented to disperse heat
    b. located in isolated areas
    c. certified by OSHA
    d. lined with lead

16. Dust-suppression equipment is maintained by _____.

    a. fuel handling workers
    b. nuclear reactor technicians
    c. line workers
    d. bulk energy brokers

17. An energy broker _____.

    a. fixes broken equipment
    b. buys and sells bulk power
    c. is licensed by OSHA
    d. runs a power plant

18. The Transportation Worker Identification Credential (TWIC) program is administered by _____.

    a. OSHA
    b. ANSI
    c. DOT
    d. TSA

19. As part of the FME program, you should _____.

    a. remove lanyards from tools
    b. inventory items entering and leaving the FME area
    c. enter the FME area by carefully stepping over the barriers
    d. carefully clean all equipment and tools after leaving the FME area

20. A typical large hydroelectric plant produces power in the realm of _____.

    a. 1,000 megawatts
    b. 500 megawatts
    c. 135 megawatts
    d. 5 megawatts

Trade Terms Quiz

Fill in the blank with the correct trade term that you learned from your study of this module.

1. A(n) _____ is a reduction of up to 25 percent in voltage from the power generating plant.

2. A(n) _____ converts a low voltage to a higher voltage or a higher voltage to a lower voltage. It is used to increase voltage to cross transmission lines and decrease it for distribution to consumers.

3. A(n) _____ is a structure used to filter fly ash from flue gases.

4. Organic matter, such as plants, used for fuel is called a(n) _____.

5. A(n) _____ is a complete loss of electrical power in an area.

6. Carbon monoxide and nitrogen oxides are examples of _____.

7. Wires that are suitable for carrying electric current are called _____.

8. The substance produced when organic matter is broken down where oxygen is not present is _____.

9. The exhaust gases produced in the furnace by the combustion process are called _____.

10. The capacity to do work is defined as _____.

11. A(n) _____ can be converted into another form.

12. A device that converts mechanical energy into electrical energy is called a(n) _____.

13. The capacity for movement is called _____.

14. A(n) _____ device converts light into DC electricity.

15. A large conducting body (such as the Earth) used as a common return for an electric circuit is called _____.

16. A(n) _____ is a device in which a nuclear fission reaction occurs under controlled conditions.

17. The type of energy that involves movement is _____.

18. The splitting of atoms of radioactive materials is called _____.

19. Material made from animals or plants is called _____.

20. A(n) _____ is made up of common transmission lines connecting two or more electric systems.

21. A device used to remove sulfurous pollutants from fossil fuel emissions is called _____.

22. The device in a power plant that provides the kinetic energy to move the electrical generator is the _____.

23. Material that radiates energy waves is classified as _____.

24. Construction costs of a generating plant built before deregulation that are not likely to be recovered are called _____.

25. Fuel removed from a reactor after use is called _____.

26. Enriched uranium is often called _____.

27. The transportation of high-voltage electricity from a generating facility is called _____.

28. The local delivery of electricity to customers is known as _____.

29. A process that breaks down coal into its chemical components by exposing it to steam and carefully controlled amounts of air or oxygen is _____.

30. A filtering device that removes particles by creating a negative charge on the particles that causes them to cling to positive plates is a(n) _____.

31. _____ is produced in the Earth from decayed animal or plant remains.

32. Electric current in which the direction of flow is reversed at set intervals is _____.

33. A material that loses resistance to the flow of electricity when cooled below a critical temperature is considered to be _____.

34. The amount of energy produced in relation to the energy input is referred to as _____.

35. A machine in which fluid energy is converted into mechanical power is a(n) _____.

Trade Terms

Alternating current	Conductors	Energy source	Nuclear fission	Spent uranium
Bag house	Distribution	Flue gases	Nuclear reactor	Stranded costs
Biogas	Efficiency	Fossil fuel	Organic matter	Superconductive
Biomass	Electrostatic	Generator	Photovoltaic	Transformer
Blackout	precipitator	Ground	Potential energy	Transmission
Brownout	Emissions	Interconnect	Prime mover	Turbine
Coal gasification	Energy	Kinetic energy	Radioactive	Yellowcake
			Scrubber	

"This is a time of tremendous growth and opportunity in the Power Generation industry. Environmental upgrades to existing fossil-fueled facilities, new nuclear generation, and the development of alternative energy sources not only require additional employees, but also employees with additional skill-sets"

Randy Fabry – Training Manager, South Carolina Electric and Gas

Ernie Griego
Arizona Public
Service Commission

"Having a national certification program that can be used as the standard, as well as management support, contributed the most to my success. Seize the opportunity and control your own career. Gain the knowledge and comfort through the modules and performance demonstrations." *Darren Marvin*

Darren Marvin
Constellation Energy

Jim Mitchem
TIC Industrial

"Power generation is a field that is not static and there is tremendous opportunity at all levels as the industry absorbs renewable energy technologies, environmental compliance enhancements, and the aging of the existing workforce.

If you understand how a facility operates you have an advantage in operations and maintenance over those without that understanding." *Jim Mitchem*

Phillip "Chip" Hammock, Jr.
Georgia Power Co.

Russell Battles
Fluor

"Individuals who choose this as their career path will always be challenged and developed; and as long as the need for electricity exists there will always be a market for their skills!" *Randy Fabry*

Randy Fabry
South Carolina
Electric and Gas

"I have taught maintenance on aircraft systems, postal sorting equipment, semiconductor manufacturing equipment and now power plant maintenance. In each of the industries I have learned new skills and knowledge that has carried over into new areas. You never stop learning during your life." *Allen Burlison*

Allen Burlison
Dunnellon
High School

Michael Hancock
Arizona Public
Service Commission,
Cholla Power Plant

"Much of my success I attribute to the sound work ethic I brought from a stint in the Air Force in wartime (Vietnam), where 16-hour workdays were the norm, and substandard work or effort was unacceptable. Never waste an opportunity to learn, whether by formal, prescribed means, or by doing the tasks others avoid." *Ernie Tomlinson*

Chris Kelley
Mechanical
Construction
Services, Inc.

Ernie Tomlinson
Salt River Project
– Coronado
Generating Station

"My career has brought me into many areas of success including opening a school which has trained over ten thousand technicians in the past 25 years." *Marcel Verroneau*

Marcel Verroneau
IMTI of NY and CT

Johnny Penrod
Arizona Public
Service Commission,
Cholla Power Plant

Trade Terms Introduced in This Module

Alternating current: Electric current in which the direction of flow is reversed at set intervals. It is measured in hertz, and 60 hertz is the standard for the United States. Frequency is also called cycles per second.

Bag house: A structure containing bags that filter flue gas to remove fly ash.

Biogas: Gas produced by the breakdown of organic matter where oxygen is not present.

Biomass: Fuel derived from recently living sources such as plants, trees, and crops, as well as livestock and human waste.

Blackout: Complete loss of electrical power in an area.

Brownout: Up to 25 percent reduction in voltage from the power-generating plant.

Coal gasification: A process that breaks down coal into its chemical components by exposing it to steam and carefully controlled amounts of air or oxygen under high temperatures and pressures.

Conductors: Wires suitable for carrying current.

Distribution: The local delivery of electricity to customers.

Efficiency: Amount of energy output from a power plant in relation to the energy input.

Electrostatic precipitator: A filtering device that removes particles by creating a negative charge on the particles that causes them to cling to positive plates.

Emissions: The byproducts of fuel combustion, such as carbon dioxide and nitrogen oxide, that are released into the atmosphere. The federal and state governments place strict limitations on the levels of emissions a facility is allowed to release to the atmosphere. Also, the radiation of energy, as radioactive energy.

Energy: The capacity to do work.

Energy source: Any source that can be converted into another type, such as electrical, mechanical, hydraulic, pneumatic, chemical, or thermal.

Flue gases: The exhaust gas produced in the furnace by the combustion process.

Fossil fuel: Fuels such as coal, oil, and natural gas that are produced in the Earth from decayed animal or plant remains. Examples of fossil fuels are coal, oil, and gas.

Generator: A device that converts mechanical energy into electrical energy.

Ground: A large conducting body (such as the Earth) used as a common return for an electric circuit and as an arbitrary zero point of potential.

Interconnect: Transmission lines connecting two or more electric systems, commonly called the grid. The grids in the United States are the Eastern Interconnection, Western Interconnection, and Texas Interconnection.

Kinetic energy: Movement.

Nuclear fission: The act of splitting atoms of radioactive materials, such as uranium, producing nuclear energy.

Nuclear reactor: A device in which a nuclear fission reaction occurs under controlled conditions to release heat.

Organic matter: Matter made from animals or plants.

Photovoltaic: A device that converts light directly into DC electricity.

Potential energy: The capacity for movement.

Prime mover: In a power plant, the device that provides the kinetic energy to move the electrical generator.

Radioactive: Radiating energy waves.

Scrubber: A device used to remove sulfurous pollutants from fossil fuel emissions.

Spent uranium: Fuel removed from a reactor after use.

Stranded costs: Construction cost of a generating plant built before deregulation that are not likely to be recovered.

Superconductive: Characteristic of certain materials that when cooled below a critical temperature lose resistance to the flow of electricity flow, thereby decreasing energy losses.

Transformer: A device that converts a low voltage to a higher voltage or a higher voltage to a lower voltage. Used to increase voltage to cross transmission lines and decrease it for distribution to consumers.

Transmission: The transportation of high-voltage electricity from a generating facility to a point where it is transformed into low voltages and distributed to customers.

Turbine: A machine in which fluid energy is converted into mechanical power.

Yellowcake: The mixture of enriched uranium oxides produced after milling uranium ore from a mine.

Additional Resources

This module is intended to present thorough resources for task training. The following references are suggested for further study. These are optional materials for continued education rather than for task training.

American Coal Ash Association website, www. acaa-usa.org.

Guide to Electrical Power Distribution Systems, Sixth Edition. Anthony J. Pansini. Lilburn, GA: The Fairmont Press, Inc.

Power Generation Technologies. Paul Breeze. Burlington, MA: Newnes.

Steam—Its Generation and Use. S.C. Stultz and J.B. Kitto. Barberton, OH: The Babcock and Wilcox Company.

Wireless Power Transfer via Strongly Coupled Magnetic Resonances. A. Kurs, K. Aristeidis, R. Moffatt, J.D. Joannopoulos, P. Fisher, and M. Soljacic. July 6, 2007. www.sciencemag.org.

Figure Credits

SAMPLES OF NCCER TRAINING CREDENTIALS

To access the Automated National Registry (ANR), visit
http://www.nccer-anr.org
and enter the number on the front of this card.

If found, drop in mailbox. Postage paid by
NCCER
3600 NW 43rd Street Building G
Gainesville FL 32606

*Showcase your credentials
Post your resume...Find jobs
http://careers.nccer.org*

NATIONAL CENTER FOR CONSTRUCTION EDUCATION AND RESEARCH

August 27, 2009

Sample Student
National Center for Construction Education and Research
3600 NW 43rd St Bldg G
Gainesville, FL 32606

Dear Sample,

On behalf of the National Center for Construction Education and Research, I congratulate you for successfully completing NCCER's Contren® Learning Series program. I also congratulate you for choosing construction as a career.

You are now a valuable member of one of our nation's largest industries. The skills you have acquired will not only enhance your career opportunities, but will help build America.

Enclosed are your credentials from the National Registry. These industry-recognized credentials give you flexibility in planning your career and ensure your achievements follow you wherever you go.

To access your training accomplishments through the Automated National Registry, follow these instructions:
1.   Go to www.nccer-anr.org.
2.   Click the "Individuals" button.
3.   Enter the NCCER card number, located on front of your wallet card or transcript, and your PIN.
       Note: The default PIN is the last four digits of your SSN. You may change your PIN after you login.
4.   First-time users will be directed to answer a few security questions upon initial login.
5.   Contact the registry department with any questions.

NCCER applauds your dedication and wishes you the best in your future endeavors.

Sincerely,

Donald E. Whyte
President, NCCER

Enc.

3600 NW 43rd St, Bldg G ○ Gainesville, FL 32606 ○ P 352.334.0911 ○ F 352.334.0932 ○ www.nccer.org

National Center for Construction Education and Research

This is to certify that

Sample Student

has fulfilled the requirements for

Your Role in the Green Environment

in NCCER's standardized training curriculum
this Tenth day of December, 2009

Donald E. Whyte

Donald E. Whyte
President

NATIONAL CENTER FOR CONSTRUCTION EDUCATION AND RESEARCH

BUILDING TOMORROW'S WORKFORCE

3600 NW 43rd St, Bldg G ○ Gainesville, FL 32606
P 352.334.0911 ○ F 352.334.0932 ○ www.nccer.org

Affiliated with the University of Florida

August 27, 2009

Official Transcript

Sample Student

National Center for Construction Education and Research

3600 NW 43rd St Bldg G

Gainesville. FL 32606

Current Employer/School:

Card #: 2781481

Course / Description		Instructor	Training Location	Date Compl.
00101	Basic Safety	Don E Whyte		1/1/2001
00102	Basic Math	Don E Whyte		1/1/2001
00103	Introduction to Hand Tools	Don E Whyte		1/1/2001
00104	Introduction to Power Tools	Don E Whyte		1/1/2001
00105	Introduction to Blueprints	Don E Whyte		1/1/2001
00106	Basic Rigging	Don E Whyte		1/1/2001
26101-02	Electrical Safety	Don E Whyte		5/1/2001
26102-02	Hand Bending	Don E Whyte		5/1/2001
26103-02	Fasteners and Anchors	Don E Whyte		5/1/2001
26104-02	Electrical Theory One	Don E Whyte		5/1/2001
26105-02	Electrical Theory Two	Don E Whyte		5/1/2001
26106-02	Electrical Test Equipment	Don E Whyte		5/1/2001
26107-02	Introduction to the National Electrical Code	Don E Whyte		5/1/2001
26108-02	Raceways, Boxes, and Fittings	Don E Whyte		5/1/2001
26109-02	Conductors	Don E Whyte		5/1/2001
26110-02	Introduction to Electrical Blueprints	Don E Whyte		5/1/2001

Donald E. Whyte
President, NCCER

Page 1

Index

A

AC. *See* alternating current (AC)
Acid rain, 1.21
Alternating current (AC), 1.1, 1.3, 1.46
American National Standards Institute (ANSI), 1.31
ANSI. *See* American National Standards Institute (ANSI)
Apprenticeship training (NCCER), 1.40
Auxiliary equipment operators, 1.28
Avedore Power Station, 1.2

B

Bag house, 1.20-1.21, 1.46
Base load, 1.38
Biogas, 1.1, 1.12, 1.46
Biomass, 1.1, 1.12-1.13, 1.39, 1.46
Blackouts, 1.22, 1.38, 1.46
Boilers, 1.2, 1.8-1.9, 1.28
Bottom ash, 1.21
Brine, 1.15
Brownout, 1.38, 1.46
Buses, 1.24

C

Careers in the power industry
    auxiliary equipment operators, 1.28
    education for, 1.28, 1.39-1.40
    energy brokers, 1.30
    fuel handling, 1.29
    job characteristics necessary for, 1.27
    job prospects, 1.28, 1.41
    laboratory analysts, 1.29
    line workers, 1.29
    maintenance workers, 1.28-1.29
    power plant operators, 1.28
Chemical energy, 1.5
Chemical safety hazards, 1.32
Chief Joseph dam, 1.13, 1.14
Chief Joseph power plant, 1.37
Clean coal technology, 1.39
Clinkers, 1.22
Coal, 1.1-1.2, 1.6
Coal-burning power plants
    chemical storage in, 1.5
    diagram, 1.4, 1.7
    efficiency of, 1.2
    overview, 1.6
    pollution control equipment
        bag house, 1.20-1.21
        combustion waste products, 1.21-1.22
        electrostatic precipitator (ESP), 1.19-1.20
        scrubbers, 1.20
        selective catalytic reduction (SCR), 1.21
        water waste, 1.21
    steam turbines in, 1.6, 1.8
    turbines in, 1.3
    waste products
        combustion waste, 1.21-1.22
        types of, 1.6
        water waste, 1.21

Coal gasification, 1.39, 1.46
Cogeneration plants, 1.2
Combined-cycle power plant, 1.10
Combustion, 1.5
Combustion turbine-generator system, 1.9-1.10
Combustion turbines, 1.9-1.10, 1.38
Combustion waste products, 1.21-1.22
Concentrated solar power (CSP), 1.18
Conductors, 1.2-1.3, 1.25, 1.46
Contren® Learning Series, 1.39-1.40
Control room operators, 1.28
Cooling ponds, 1.11, 1.21
Craft instructor, 1.40
CSP. *See* concentrated solar power (CSP)
Current wars, 1.1

D

DC. *See* direct current (DC)
Department of Homeland Security, 1.32
Department of Labor, 1.40
Department of Transportation (DOT), 1.31-1.32
Direct current (DC), 1.1
Distribution, 1.46
Distribution systems, underground, 1.24, 1.26. *See also* transmission and distribution systems
DOT. *See* Department of Transportation (DOT)
Dry geothermal energy, 1.16-1.17

E

Edison, Thomas, 1.1
Education for power-industry workers
    continuing, 1.28
    NCCER, 1.39-1.40
Efficiency
    of biomass, 1.13
    coal-burning power plants, 1.2
    defined, 1.2, 1.46
    hydropower, 1.13
    recirculation systems for, 1.8
Electrical energy, 1.5
Electrical safety, 1.31, 1.32
Electric power
    costs, construction and operating, 1.37
    generation, 1.1-1.3
    history of, 1.1
    hydrogen fuel cells, 1.39
    reliability, 1.37-1.38
    summary, 1.41
    transmission and distribution systems
        complexity in, 1.37-1.38
        federal regulatory agency, 1.37
        historically, 1.22
        illustrated, 1.5
        interconnects, 1.22
        laws regulating, 1.36
        overview, 1.3-1.4
        power-handling capability, 1.38
        subsystems, 1.22-1.26

Electrostatic precipitator (ESP), 1.19-1.20, 1.46
Emissions, 1.6, 1.11, 1.46
Energy
    chemical, 1.5
    defined, 1.1, 1.46
    electrical, 1.5
    kinetic, 1.4, 1.46
    mechanical, 1.5
    potential, 1.4-1.5, 1.46
    renewable/nonrenewable, 1.5
    thermal, 1.5
Energy brokers, 1.30
Energy conservation principle, 1.4
Energy conversion, four-way, 1.5
Energy Policy Act (EPACT), 1.36
Energy sources
    defined, 1.1, 1.46
    future of, 1.38-1.39
    hydrogen, 1.39
    renewable
        biomass fuel, 1.12-1.13
        defined, 1.5
        geothermal, 1.15-1.17
        introduction, 1.12
        solar, 1.18
        water (hydropower), 1.13-1.14
        wind, 1.17-1.18
Environmental Protection Agency (EPA), 1.31
Environmental protections
    dam construction, 1.14
    EPA oversight, 1.31
    future of, 1.38-1.39
    pollution control equipment
        bag house, 1.20-1.21, 1.46
        combustion waste products, 1.21-1.22
        electrostatic precipitator (ESP), 1.19-1.20
        scrubbers, 1.20, 1.46
        selective catalytic reduction, 1.21
        water waste, 1.21
EPA. See Environmental Protection Agency (EPA)
EPACT. See Energy Policy Act (EPACT)
ESP. See electrostatic precipitator (ESP)
Ethanol, 1.12

F
Federal Aviation Administration, 1.31
Federal Energy Regulatory Commission (FERC), 1.37
Federal Highway Administration, 1.31
FERC. See Federal Energy Regulatory Commission (FERC)
Flue gases, 1.19, 1.46
Fly ash, 1.19-1.20, 1.21
FME. See foreign material exclusion (FME)
Foreign material exclusion (FME), 1.33-1.34
Fossil-fuel power plants, overview, 1.6-1.10
Fossil fuels, 1.5, 1.6, 1.46
Fuel cells, 1.39
Fuel handling, 1.29
Full power, 1.38

G
General Electric, 1.1
Generator, 1.1, 1.2-1.3, 1.46
Geothermal energy, 1.15-1.17
Ground, 1.25, 1.46
Ground fault, 1.35
Grounding grid, 1.25-1.26, 1.34-1.35
Grounding safety practices, 1.34-1.35
Gypsum, 1.20

H
Heliostats, 1.18
Homeland Security Department, U.S., 1.32
Hoover Dam, 1.3
Hopper ash, 1.21
Hydroblasting, 1.22
Hydroelectric plants, 1.3, 1.37
Hydrogen fuel cells, 1.39
Hydropower energy, 1.13-1.14

I
IEEE. See Institute of Electrical and Electronics Engineers (IEEE)
Institute of Electrical and Electronics Engineers (IEEE), 1.31
Interconnect, 1.22, 1.46
Inverter, 1.18

K
Kinetic energy, 1.4, 1.46

L
Laboratory analyst, 1.29
Labor Department, U.S., 1.40
Line workers, 1.29
Load balancing, 1.22
Load center, 1.25

M
Maintenance workers, 1.28-1.29
Maritime transportation system, 1.32
Mechanical energy, 1.5
Mechanical safety hazards, 1.32
Methane, 1.6, 1.12
Mine Safety and Health Administration (MSHA), 1.31
Mining industry oversight, 1.31
MSHA. See Mine Safety and Health Administration (MSHA)

N
National Apprenticeship System, 1.40
National Center for Construction Education and Research (NCCER)
    apprenticeship training, 1.40
    standardized training programs, 1.39-1.40
    training credentials, 1.49-1.52
    youth training and apprenticeship programs, 1.40
National Center for Construction Education and Research (NCCER) Accreditation Guidelines, 1.39
National Center for Construction Education and Research (NCCER) Accredited Training Sponsors, 1.39-1.40
National Center for Construction Education and Research (NCCER) National Registry, 1.39
National Electrical Safety Code, IEEE/ANSI C2 (ANSI), 1.31
National Electric Reliability Corporation (NERC), 1.22, 1.37
National Fire Protection Association (NFPA), 1.31
National power grid, 1.41
Natural gas, 1.6
Natural gas power plants, 1.8, 1.10
NCCER. See National Center for Construction Education and Research (NCCER)
NERC. See National Electric Reliability Corporation (NERC)
NFPA. See National Fire Protection Association (NFPA)
Nitrogen oxide (NO_x) emissions, 1.21
Nonrenewable energy sources, 1.5
Non-selective SCR (NSCR) technology, 1.21
North American Electric Reliability Corporation (NERC), 1.22
NRC. See Nuclear Regulatory Commission (NRC)
NSCR. See non-selective SCR (NSCR) technology
Nuclear fission, 1.11, 1.46
Nuclear power, 1.11

Nuclear power plant operators, 1.28
Nuclear power plants, 1.11-1.12, 1.32, 1.38
Nuclear reactor, 1.11, 1.46
Nuclear Regulatory Commission (NRC), 1.28, 1.32

O

Occupational Safety and Health Administration (OSHA), 1.30-1.31, 1.35
Office of Apprenticeship (Labor Department), 1.40
Oil, 1.6
Oil-burning power plants, 1.8-1.9
OJL. *See* on-the-job learning (OJL)
Old Faithful, 1.15
On-the-job learning (OJL), 1.28, 1.40
Organic matter, 1.6, 1.46
OSHA. *See* Occupational Safety and Health Administration (OSHA)

P

Peakers, 1.38
Peak load, 1.38
Personal protective equipment (PPE), 1.33
Personal protective grounding (PPG), 1.35
Photovoltaic (PV), 1.18, 1.46
Plutonium, 1.11
Pollution-control equipment. *See also* environmental protections
    bag house, 1.20-1.21, 1.46
    combustion waste products, 1.21-1.22
    electrostatic precipitator (ESP), 1.19-1.20
    scrubbers, 1.20, 1.46
    selective catalytic reduction, 1.21
    water waste, 1.21
Potential energy, 1.4-1.5, 1.46
Power grid
    controlling voltage levels, 1.23
    development of, 1.22
    national, 1.41
    purpose, 1.3-1.4
    smart-grid technology, 1.39
Power industry. *See also* careers in the power industry
    economics, 1.36-1.38
    future of
        environmental considerations, 1.38-1.39
        technological advances, 1.39
    regulatory oversight, 1.36-1.37
    summary, 1.41
Power outages, 1.22, 1.38, 1.46
Power plants. *See also* coal-burning power plants
    coal gasification, 1.39
    combined-cycle, 1.10
    efficiency of, 1.2, 1.10
    fossil-fuel, 1.6-1.10
    lower capacity, 1.9
    reliability, 1.37-1.38
    simple-cycle combustion, 1.10
    typical production, 1.37
Power plant workers
    auxiliary equipment operators, 1.28
    education for, 1.28, 1.39-1.40
    energy brokers, 1.30
    fuel handlers, 1.29
    laboratory analysts, 1.29
    line workers, 1.29
    maintenance workers, 1.28-1.29
    nuclear power plant operators, 1.28
    power plant operators, 1.28
    safety hazards specific to, 1.32

Transportation Worker Identification Credential (TWIC), 1.32
Power transmission switchyard, 1.24
PPE. *See* personal protective equipment (PPE)
PPG. *See* personal protective grounding (PPG)
Pressure safety hazards, 1.32
Prime mover, 1.6, 1.46
PV. *See* photovoltaic (PV)

R

Radiation, 1.11
Radiation decay time, 1.11
Radioactive, 1.11, 1.46
Recirculation process, 1.8
Related technical instruction (RTI), 1.40
Renewable energy sources
    biomass fuel, 1.12-1.13
    defined, 1.5
    future of, 1.38-1.39
    geothermal, 1.15-1.17
    introduction, 1.12
    solar, 1.18
    water (hydropower), 1.13-1.14
    wind, 1.17-1.18
Rotor, 1.2-1.3, 1.8
RTI. *See* related technical instruction (RTI)
Rural Electrification Program, 1.14

S

SAC. *See* state apprenticeship councils (SAC)
Safety agencies
    American National Standards Institute (ANSI), 1.31
    Department of Transportation (DOT), 1.31-1.32
    Environmental Protection Agency (EPA), 1.31
    Mine Safety and Health Administration (MSHA), 1.31
    Nuclear Regulatory Commission (NRC), 1.32
    Occupational Safety and Health Administration (OSHA), 1.30-1.31, 1.35
    Transportation Security Administration (TSA), 1.32
Safety education, 1.28
Safety hazards, power-industry specific, 1.32
Safety practices
    foreign material exclusion (FME), 1.33-1.34
    grounding, 1.34-1.35
    policies and procedures, 1.32-1.33
Safety regulations
    changes to, suggesting, 1.31
    following, 1.31
    importance of, 1.30
    regulatory oversight, 1.30-1.32, 1.35
Scrubber, 1.20, 1.46
Selective catalytic reduction, 1.21
Simple-cycle combustion turbine, 1.10
Slag, 1.21-1.22
Smart-grid technology, 1.39
Smog, 1.21
Solar energy, 1.18
Spent uranium, 1.11, 1.46
Standards system, 1.31
Stanley, William, Jr., 1.1
State apprenticeship councils (SAC), 1.40
Steam turbines, 1.6, 1.8-1.9, 1.15
Stranded costs, 1.37, 1.46
Sulfur dioxide (SO$_2$), 1.19-1.20
Superconductive, 1.39, 1.46
Switchyards, 1.24-1.26

T

Technology, advances in, 1.39
Temporary protective grounds (TPGs), 1.35
Tesla, Nikola, 1.1, 1.39
Thermal energy, 1.5
Thermal safety hazards, 1.32
Thermodynamics, first law of, 1.4
TPGs. *See* temporary protective grounds (TPGs)
Transformer
    defined, 1.3, 1.47
    pad- and pole-mounted, 1.26
    step-up/step-down, 1.23
Transmission, 1.1, 1.22, 1.47
Transmission and distribution systems
    historically, 1.22
    laws regulating, 1.36
    overview, 1.3-1.4
    power-handling capability, 1.38
    substations and switchyards, 1.24-1.26
    subsystems, 1.22-1.26
    superconductive, 1.39
    technological advances, 1.39
Transportation Department (DOT), 1.31-1.32
Transportation Security Administration (TSA), 1.32
Transportation system oversight
    Department of Homeland Security, 1.32
    Department of Transportation (DOT), 1.31-1.32
Transportation Worker Identification Credential (TWIC), 1.32
TSA. *See* Transportation Security Administration (TSA)
Turbine-generator combination, 1.3
Turbines
    combustion, 1.9-1.10, 1.38
    defined, 1.3, 1.47
    geothermal energy, 1.15
    hydroelectric, 1.13, 1.14
    steam, 1.5-1.6, 1.8-1.10
TWIC. *See* Transportation Worker Identification Credential (TWIC)

U

Underground distribution systems, 1.24, 1.26
Uranium, 1.11, 1.31, 1.46
Utility companies, 1.36

W

Waste heat, 1.8
Waste products
    coal-burning power plants, 1.6, 1.21-1.22
    coal gasification, 1.39
    combustion waste, 1.21-1.22
    heat, 1.8
    hydrogen fuel cells, 1.39
    nuclear power plants, 1.11
    oil-burning power plants, 1.9
    pollution-control equipment
        bag house, 1.20-1.21
        combustion waste products, 1.21-1.22
        electrostatic precipitator (ESP), 1.19-1.20
        scrubbers, 1.20
        selective catalytic reduction, 1.21
        water waste, 1.21
    water waste, 1.11, 1.21
Water energy, 1.13-1.14
Water waste, 1.11, 1.21
Westinghouse, George, 1.1
Wet geothermal energy, 1.15-1.16
Wet scrubber, 1.20
Wind energy, 1.17-1.18
Work, 1.5

Y

Yellowcake, 1.11, 1.47
Youth training and apprenticeship programs (NCCER), 1.40

NCCER makes every effort to keep these textbooks up-to-date and free of technical errors. We appreciate your help in this process. If you have an idea for improving this textbook, or if you find an error, a typographical mistake, or an inaccuracy in NCCER's Contren® textbooks, please write us, using this form or a photocopy. Be sure to include the exact module number, page number, a detailed description, and the correction, if applicable. Your input will be brought to the attention of the Technical Review Committee. Thank you for your assistance.

Instructors – If you found that additional materials were necessary in order to teach this module effectively, please let us know so that we may include them in the Equipment/Materials list in the Annotated Instructor's Guide.

Write: Product Development and Revision
National Center for Construction Education and Research
3600 NW 43rd St, Bldg G, Gainesville, FL 32606

Fax: 352-334-0932

E-mail: curriculum@nccer.org

Craft _____ Module Name _____

Copyright Date _____ Module Number _____ Page Number(s) _____

Description _____

_____

_____

_____

(Optional) Correction _____

_____

_____

(Optional) Your Name and Address _____

_____

_____